Visual Environmental Communication

In 2008, the editors published a well-cited journal paper arguing that while scholarly work on media representations of environmental issues had made substantial progress in textual analysis there had been much less work done on visual representations. This is surprising given the increasingly visual nature of media and communication, and in light of emerging evidence that the environment is visualized through the use of increasingly symbolic and iconic images.

Addressing these matters, this volume marks out the present state of the field and contains chapters that represent fresh and exciting high quality scholarly work now emerging on visual environmental communication. These include a range of fascinating and often alarming topics which draw on a variety of methods and forms of visual communication. The book demonstrates that research needs to think much more widely about what we mean by the 'visual' which plays a massive yet under-researched role in the politics and ideology of public understanding and misunderstanding of the environment and environmental problems.

The book is of relevance to students and researchers in media and communication studies, cultural studies, film and visual studies, geography, sociology, politics and other disciplines with an interest in the politics of visual environmental communication.

This book was previously published as a special issue of *Environmental Communication*.

Anders Hansen is a Senior Lecturer in the Department of Media and Communication, University of Leicester, UK. He has published extensively on media and the environment, including *Environment, Media and Communication* and the forthcoming *The Routledge Handbook of Environment and Communication* (co-edited with Robert Cox). He is founder and ex-Chair of the IAMCR Environment, Science and Risk Communication Group, founding member and Secretary to the International Environmental Communication Association (IECA), and Associate Editor for *Environmental Communication*.

David Machin is Professor of Media and Communication at Örebro University, Sweden. He has published extensively on visual communication and Multimodal Critical Discourse Analysis, such as *Introduction to Multimodal Analysis*, *Global Media Discourse* and *The Language of War Monuments* and a recent multidisciplinary edited collection called *Visual Communication*. He is co-editor for the international peer reviewed journals *Journal of Language and Politics* and *Social Semiotics*.

Visual Environmental Communication

Edited by
Anders Hansen and David Machin

LONDON AND NEW YORK

First published 2015 by Routledge

2 Park Square, Milton Park, Abingdon, Oxon, OX14 4RN
605 Third Avenue, New York, NY 10017

Routledge is an imprint of the Taylor & Francis Group, an informa business

First issued in paperback 2020

Copyright © 2015 Taylor & Francis

All rights reserved. No part of this book may be reprinted or reproduced or utilised in any form or by any electronic, mechanical, or other means, now known or hereafter invented, including photocopying and recording, or in any information storage or retrieval system, without permission in writing from the publishers.

Notice:
Product or corporate names may be trademarks or registered trademarks, and are used only for identification and explanation without intent to infringe.

British Library Cataloguing in Publication Data
A catalogue record for this book is available from the British Library

ISBN 13: 978-1-138-80375-6 (hbk)
ISBN 13: 978-0-367-73871-6 (pbk)

Typeset in Minion
by RefineCatch Limited, Bungay, Suffolk

Publisher's Note
The publisher accepts responsibility for any inconsistencies that may have arisen during the conversion of this book from journal articles to book chapters, namely the possible inclusion of journal terminology.

Disclaimer
Every effort has been made to contact copyright holders for their permission to reprint material in this book. The publishers would be grateful to hear from any copyright holder who is not here acknowledged and will undertake to rectify any errors or omissions in future editions of this book.

Contents

Citation Information   vii

1. Introduction: Researching Visual Environmental Communication   1
   Anders Hansen & David Machin

2. Visualizing the Chesapeake Bay Watershed Debate   19
   Elizabeth Anne Gervais Schwarz

3. Imaging Toxins   41
   Jennifer Peeples

4. Selling Nature in a Resource-Based Economy: Romantic/Extractive Gazes and Alberta's Bituminous Sands   61
   Geo Takach

5. "Single-minded, compelling, and unique": Visual Communications, Landscape, and the Calculated Aesthetic of Place Branding   79
   Nicole Porter

6. The Nature of *Time*: How the Covers of the World's Most Widely Read Weekly News Magazine Visualize Environmental Affairs   103
   Mark S. Meisner & Bruno Takahashi

7. Sporting Nature(s): Wildness, the Primitive, and Naturalizing Imagery in MMA and Sports Advertisements   125
   Matthew P. Ferrari

8. Mobilizing Artists: *Green Patriot Posters*, Visual Metaphors, and Climate Change Activism   145
   Brian Cozen

Index   163

Citation Information

The chapters in this book were originally published in *Environmental Communication*, volume 7, issue 2 (2013). When citing this material, please use the original page numbering for each article, as follows:

Chapter 1
Editors' Introduction: Researching Visual Environmental Communication
Anders Hansen & David Machin
Environmental Communication, volume 7, issue 2 (2013) pp. 151–168

Chapter 2
Visualizing the Chesapeake Bay Watershed Debate
Elizabeth Anne Gervais Schwarz
Environmental Communication, volume 7, issue 2 (2013) pp. 169–190

Chapter 3
Imaging Toxins
Jennifer Peeples
Environmental Communication, volume 7, issue 2 (2013) pp. 191–210

Chapter 5
"Single-minded, compelling, and unique": Visual Communications, Landscape, and the Calculated Aesthetic of Place Branding
Nicole Porter
Environmental Communication, volume 7, issue 2 (2013) pp. 231–254

Chapter 6
The Nature of Time*: How the Covers of the World's Most Widely Read Weekly News Magazine Visualize Environmental Affairs*
Mark S. Meisner & Bruno Takahashi
Environmental Communication, volume 7, issue 2 (2013) pp. 255–276

CITATION INFORMATION

Chapter 7
Sporting Nature(s): Wildness, the Primitive, and Naturalizing Imagery in MMA and Sports Advertisements
Matthew P. Ferrari
Environmental Communication, volume 7, issue 2 (2013) pp. 277–296

Chapter 8
Mobilizing Artists: Green Patriot Posters, *Visual Metaphors, and Climate Change Activism*
Brian Cozen
Environmental Communication, volume 7, issue 2 (2013) pp. 297–314

Please direct any queries you may have about the citations to
clsuk.permissions@cengage.com

INTRODUCTION
Researching Visual Environmental Communication

Anders Hansen & David Machin

In 2008, we published a journal paper arguing that while scholarly work on media representations of environmental issues had made substantial progress in textual analysis, there had been much less work on visual representations. This special edition has a number of aims in this respect. It seeks to mark out where there has been progress since 2008, and the papers in this collection represent some of the fresh and exciting high quality scholarly work now emerging on an expanding number of topics and using different methods. We argue that we need to think more openly about what we mean by "the visual." We begin by placing research into visual representations of the environment into the wider trajectory of visual studies research. We then proceed to review key trends in visual environmental communication research and to delineate core dimensions, contexts and sites of visual analysis.

In 2008, we published a journal paper arguing that, while scholarly work on media representations of environmental issues had made substantial progress in textual analysis (e.g., Boykoff, 2007, 2008; Carvalho & Burgess, 2005; Shanahan & McComas, 1999), much less work had been done on visual representations. This special edition has a number of aims in this respect. It seeks to mark out where there has been progress since that time. The papers in this collection represent some of the fresh and exciting high quality scholarly work now emerging on an expanding number of topics and using different methods. We want to take this opportunity to establish where we

---

Anders Hansen is a Senior Lecturer in the Department of Media and Communication, University of Leicester, UK; David Machin is Reader in Journalism, School of Arts, Brunel University, UK.

are now, what questions we should now be asking, and in what directions we might fruitfully move. We show for one thing that we need to think more openly as regards what we mean by "the visual." We begin to do this by placing the research of visual representations of the environment into the wider trajectory of visual studies research. This provides some clearer indications of what is at stake specifically as regards "the visual." We then proceed to review key trends in visual environmental communication research and to outline some of the key areas and challenges for moving forward.

Researching Visual Communication

A number of scholars have observed a broader growth in interest in visual communication across academic disciplines (Pauwels, 2012; Rose, 2012). On the one hand, this increase represents a growing acknowledgment of the important role played by visual communication. On the other, it also represents a sense that scholars are coming together to share knowledge of visual communication and also looking outward to other disciplines for new methods and theories. Pauwels (2012) points out that there has been a tendency in visual research for different fields to reinvent the wheel as they operate in their own isolated networks. Pauwels presents his own edited collection motivated to provide a more integrated set of visual methods. And certainly we would see part of the role of this special issue to showcase the different kinds of methods and approaches that scholars from different fields bring to what we term as a set of core concerns.

This increase in interest in visual communication can be traced to a number of influences, and considering these and their later trajectory is important in allowing us to identify what kind of path we may need to follow in the study of visual representations of the environment. The idea of the study of visual communication in the areas of media and culture can be placed from the 1970s with the interest in the ideas of Marxist thinkers, such as Adorno and Horkheimer, and then the 1980s with the influence particularly of the Birmingham School of Cultural Studies. The work of people like Stuart Hall (1977) had a considerable influence, particularly in the UK, pointing to the importance of the visual in culture, how we needed to think of culture not simply as things like art and theater, but in an anthropological sense, as a set of practices and beliefs that formed a set of values, attitudes and ideas about how the world worked, and what was important. What is culture could be thought of as what is accepted as common sense knowledge of how things are, of who we are, and of how the world works.

Such a notion of culture led to a switch in the idea of culture as a reflection of society to one where it could cause and shape society. And much research in media and cultural studies belong to this kind of tradition, drawing out the ideologies realized in images. And to some extent the activity of researching visual representations of the environment is of this nature, as we seek to unpack the kinds of cultural baggage and ideologies that shape our thoughts about the planet, about nature and the nature of the threats it faces.

Further, this view of culture means that we look to the nature of the industries where representations are reproduced. Here we seek to understand the production and the economics that are behind production, such as the way in which individual media outlets are owned by globally operating conglomerates often interlinked with wider corporate and financial institutions (McChesney, 2004), how advertisements are the engines of mass media (Jhally, 1991), and how certain kinds of media products are financed and distributed (McDonald & Wasko, 2007).

This tradition of research into visual culture also placed "seeing" as one important critical point of analysis. Seeing is a result of the cultures in which we have grown up and live. Representations in news, photography, television and films are part of the manner in which culture shapes our ways of seeing. There have been a number of works on this from Berger's seminal *Ways of seeing* (1972). What is of interest here is why particular viewers look in different ways and how images invite different kinds of views. As an example from art history, Berger shows that women in oil paintings were depicted as passive, naked, vain and sexually alluring, pointing to the fact that they were designed to be looked at by men (1972, p. 54). Mulvey's (1975) well-cited study on the "male gaze" in cinema argues that movies present a masculine view of female passivity and sexuality. Seeing has also been thought of in terms of the ways in which different kinds of viewers in terms of gender, social class, age, will interpret what they view (Hall, 1973/1980). These ideas call us to think about how visual representations of the environment point to cultures of viewing; for example, the viewing of landscapes through a romantic gaze, seeing them as representing some pristine and innocent view of nature, or seeing the planet in terms of resources that can be exploited.

This sense of different kinds of seeing reminds us that we need to understand how different kinds of people see images of the environment, whether produced by marketing companies to show they are carbon friendly, and the clichés produced by news programs, or those produced by environmental campaigners. We need to ask how these appeal to different kinds of viewing sensibilities.

All of this emphasis on the visual was especially important, as scholars like Jenks (1995) point out, as Western culture has come to equate seeing with knowing, even if this seeing is essentially culturally loaded. Central to this line of research was the perceived documentary role of the photograph. Early uses of the photograph were celebrated for their power to document and bear witness, but scholars working in Visual Studies began to question this power. Authors like Sontag (2004) pointed to the way in which iconic images represent and replace complex processes. Others made the observation that many images presented to the public claiming to document and bear witness, rather lean on well-trodden themes that are consonant and chime with existing sets of values (Cottle, 2009).

In the late 1980s, Baudrillard (1988) warned that due to such images we had ceased to be able to make a distinction between the real and the represented; such was the nature of living in a world with such a flow of images. This echoed the earlier concerns of Barthes (1973), who argued that the image can allow a predominance of certain kinds of mass disseminated myths that tend to support the interests of the powerful in society. And many scholars, influenced also by the ideas of Foucault (1972), began to think about the way that the interests of the powerful can be naturalized through the dissemination of particular kinds of images that foster specific kinds of ideas, values and identities that favor the world of corporate capitalism. In our own work (Hansen & Machin, 2008), we have looked at the way that commercial image archives supply images to media across the planet that are designed precisely to point to the kinds of clichés and myths writers like Baudrillard and Foucault had in mind.

Finally, much scholarly work has gone into *how* the visual communicates. Different kinds of interests and concerns are found across different fields. In art, an analysis might approach a painting in terms of use of perspective, of conventions of representing landscapes, scenes and figures and lighting (Panofsky, 1972). In photography, we can look at focus, angle and point of view, in other words composition and the perspective set up for the viewer (Barthes, 1977). For movies, we can add dimensions like editing, cuts, continuity and slow motion (Monaco, 2009) or look at them in terms of the way that they construct the world through the film narrative form (Bordwell & Thompson, 2008; Metz, 1974). Analyzing all of these can help to reveal how representations are being shaped and manipulated for the viewer.

Of course, the realm of how images work has been one core focus of semiotics, although in Anglo American scholarship it is primarily the work of Barthes (1973, 1977) that has been more widely used to look at the way elements and styles in images can be used to connote wider meanings and values that may not be otherwise overtly stated. More recently the work of linguists, such as Kress and van Leeuwen (1996/2006) in their classic work *Reading Images*, has sought to provide more systematic tools for relating matters in visual composition to underlying ideologies.

While these approaches to visual culture appear on the face of it pretty comprehensive, authors such as Elkins (2003) have been critical of the rather narrow definitions of the visual used in the field with tendency to focus on television, film and photographs. But visual culture infuses our lives in so many more ways, each in its own way communicating ideas, values and identities and providing different kinds of seeing. Visual communication can be done through the clothes we wear, through gesture, through the way we layout our homes, through the way houses are built to suggest austerity and conformity or their opposites. For the likes of Elkins, it is this wider sense of visual communication that we must embrace.

So, to what extent can research into visual representations of the environment expand its sense of the visual? There is room here to think in terms of the political economy of these representations, the ideologies they carry and the kinds of culturally loaded viewing that they carry, along with the way that different kinds of viewers experience them.

Studies of Visual Representations of the Environment

In (2010), Hansen pointed to the rise since the early 1960s of a familiar, yet superficial, public vocabulary and discourse on a wide range of "problems" and "issues" seen as associated with our natural environment. Key terms and labels characterizing this vocabulary have become familiar through sheer repetition in public media and communication. But while much research has focused on the textual, rhetorical and linguistic construction of the public vocabulary on the environment, much less attention has been given to its visual articulation and construction. Yet, the public vocabulary on the environment is to a large extent a visual vocabulary, and the very same kinds of issues that have emerged in visual studies more broadly are of equal relevance here. These include the problematic tendency of conflating "seeing" with "knowing or understanding" and the culturally conditioned notion that images "document reality" in a less mediated, constructed or manipulated way than other types of communication.

The invisibility and slow development of many environmental problems pose particular difficulties for their news construction and communication generally, and for their visual representation specifically. As many scholars have pointed out (e.g., Adam, 1998; Cox, 2013; Doyle, 2007; Hannigan, 2006), the difficulty of providing simple visual representations of causes and consequences poses a challenge to their visualization and more widely to the acceptance of their very existence. In an early study of environmental news, Schoenfeld, Meier, and Griffin (1979) noted that one of the key challenges for news coverage of environmental issues was the mismatch between the demands of a rapid news-cycle and the often long-term and slow-evolving pace of many environmental problems. But the problem extends significantly further than a mismatch of cycles: many environmental problems are just not that visible and the substances that cause them may themselves be invisible or innocuous-looking (Peeples, 2011). These are characteristics that make environmental problems and their visual representation more open – than, for example, their representation in text/language – to interpretation and indeed to ideological manipulation.

It has long been recognized that perceptions/images of nature are socially, politically and culturally constructed (Macnaghten & Urry, 1998; Soper, 1995; Urry, 1992; Williams, 1973, 1976/1983). Historically specific constructions and visual representations of nature are used – and exploited ideologically – to inform everything from public debate about genetics (Hansen, 2006), advertising and marketing of products, places and ideas (Ahern, Bortree, & Smith, 2012; Hansen, 2002; Svoboda, 2011; Williamson, 1978), to the marketing of tourist destinations (Todd, 2010; Urry, 1992), and to television documentaries or films (Bousé, 2000; Ferreira, 2004; Mitman, 1999; Rust, Monani, & Cubitt, 2013; Wall, 1999). For most of these studies, however, visual analysis has not been the central focus. It is only in recent years that studies focusing on the visual construction of the environment have begun to appear in greater numbers.

A number of studies (e.g., Cottle 2000; Doyle 2007; Hansen & Machin 2008; Lester & Cottle 2009; Linder 2006; Szerszynski, Urry, & Myers, 2000) have been important earlier exceptions to this pattern and comprise the first wave of research in the field. These studies all draw considerably on the semiotics of Roland Barthes, while Linder extends this with the more recent work on visual semiotics by Kress and van Leeuwen (1996/2006) and also on the tradition of Critical Discourse Analysis.

In their analysis of global images on British television, Szerszynski et al. (2000) found extensive use across a range of genres of – often abstracted or decontextualized – "global imagery" in the form of *globes, representative environments* and *representative people*. Otherwise ordinary images were framed as global, but usually served as general background to the main message in a way that Szerszynski et al characterize as "banal globalism" (p. 106).

Cottle (2000), in his analysis of television news coverage of the environment, revealed a visual emphasis drawing on romanticized views of nature and representing the environment as spectacle, landscape and "under threat." Rather than interrogating specifics of particular processes of threats to the environment, television deploys, as Cottle argues, "a crafted succession of iconic and symbolic images" which have become part of an "almost standardized visual 'lexicon' deployed in the representation of environmental disaster stories…." (p. 41).

The indication from these and other studies (including Hansen & Machin, 2008) is that television and other media visualize the environment through the use of increasingly decontextualized global, symbolic and iconic images rather than those which are recognizable because of their geographic/historical or social/cultural anchoring. As we have argued previously (Hansen & Machin, 2008), an ideological consequence of this form of decontextualization is a visual disconnect from concrete processes, such as global capitalism and consumerism.

Of course, many environmental issues, as we have noted above, are relatively invisible and difficult to depict visually (Peeples, 2011). Importantly, environmental images do not acquire iconic or representative status by themselves. This requires visual signification work in much the same way as environmental issues, and only become issues for public and political concern through the public claims-making activities of scientists, pressure groups, governments and others. Linder's (2006) analysis tracks some of these important signification processes. He shows how global warming signs/visualization, originating in the scientific/regulatory/political discourses of environmental groups and governments, are appropriated, and sometimes inverted, by advertisers and end up in the service of the promotion of consumption.

Linder provides important comments on some of the key signification processes involved in the visualization of the environment, particularly the *decontextualization* and *aestheticization* of landscapes or physical settings and the use of imagery, which *resonates with deeper cultural discourses* or myths of for example, unspoilt wilderness as national heritage. In her study of Greenpeace's use of visuals in their climate campaigns, Doyle (2007) similarly alludes to the important ideological implications of inscribing campaign images esthetically into the romantic tradition of landscape images.

A further important dimension emerging from studies of climate change and other environmental visualization, such as those just mentioned is the temporal/processual dimension of signification, the notion that visual signification is anything but static, and that the elevation of particular images to iconic status as images *representing* a particular meaning, such as climate change or environmental devastation or (perhaps more obliquely) threatened environments is an ongoing process drawing on, what Linder (2006) aptly refers to as "an extensive collection of semiotic resources" (p. 130) and involving "a substantial amount of appropriation and pastiche between them, as they exploit newly established signs in novel variations" (pp. 129–130). We shall return to this in our discussion of Historical Context below.

In our own study (Hansen & Machin, 2008), we looked at the way that commercial image suppliers, such as Getty Images provide media with cheap and attractive images that have semiotic flexibility – and adaptability – afforded by decontextualization and the absence of the recognizable identifiers crucial to realist/documentary photography. It is precisely this semiotic openness that makes the images ideal for a global media market and for hard-pressed news organizations, which have neither the time nor the resources to deploy expensive camera-crews to collect documentary-style, factual, real images to satisfy increasingly visual-focused media and a visually increasingly hungry audience.

Lester and Cottle (2009) make similar points as regards the symbolism of climate change, here in the case of television news: how a single smoldering tree stump can symbolize the whole process of deforestation, how a single billowing chimney can represent the whole of global industrial production. Such images, they point out, while largely symbolic, have been a means to make climate change into the perceived global crisis that we now face, and have played an important role in displacing the challenges of skeptics. These authors also draw on the case made by Szerszynski et al. (2000), who argue that routine images of the globe point to an un articulated sense of common citizenship, which suggests that we all belong to the same planet, and where celebrities are seen to stand up and speak for the whole of the human race (Lester & Cottle, 2009, p. 923).

In summary, this literature points to the way in which news and advertising images draw on a stock range of symbolic images and ones that draw on a romanticized view of nature. Visual representations of the environment tend to be decontextualized and aestheticized in ways that enhance their flexible and versatile use across different genres of communication, while also affording the basis for flexible new significations, as well as ones that are firmly anchored in culturally deep-seated/resonant discourses on nature and the environment. On the one hand, the invisibility and slow development of many environmental problems provides an obstacle to their realist representation. And on the other, news organizations tend to lean toward well-trodden frames of reference to make issues more easily recognizable to audiences. In the case of advertising and marketing, it is clear that certain kinds of representations are favored that allow products and services to be loaded with a moral sense of connecting with nature and the environment. In all cases, representations appear to favor individual responses to environmental problems rather than those that call for major structural changes in terms of the way in which we organize our societies and the resource greedy nature of capitalism.

In the earlier section of this introduction we looked at some of the trajectories of research in visual studies: the ownership and production of images, the ideas and values these carry, how we have culturally tutored ways of looking that may be different across different kinds of people and how we can look at the conventions and structures of visual representation themselves. But how close are we to providing a comprehensive picture of these dimensions of the visual regarding representations of the environment? And how far can we avoid Elkin's (2003) criticism of having a very limited view of what comprises the visual? With this question in mind, we now turn, in the remainder of the paper, to outline what we see as the core foci, contexts and sites for the analysis of visual environmental communication.

Interacting Sign Systems

Semiology, of course, means the "science of signs," and this perhaps signals one of the most important contributions of founding father Ferdinand de Saussure's argument that language (and other sign systems) should be studied as meaning creating systems. But perhaps the most significant insight in this context is Barthes' argument (1973, 1977) that our communicative environment consists of several important sign systems, and that meaning is created both within each of these and significantly through their interaction. Of principal interest to studies of visual environmental communication has been the interaction of images and linguistic text.

For example, in their analysis of climate change imagery in the British press, Smith and Joffe (2009) argue that images provide a more narrow focus and emphasis to the (more complex) articulations of the written text in newspaper reporting on climate change. By contrast, DiFrancesco and Young (2011) in a rigorous analysis of image-language interactions in "the visual construction of global warming in Canadian national print media" conclude that images and text are pointing in surprisingly different directions: it is neither the case that images are driving the content of articles, nor the other way round. Rather, they argue, "it appears that in many cases journalists and editors are attaching images *post-facto* to articles that tend to be morally or emotionally edgy, regardless of the content of those images" (p. 532). In an insightful analysis of global warming documentaries, Mellor (2009) likewise notes the problematic relationship between talking heads statements and the extensive use of wall-paper shots – generalized, decontextualized images of the environment – that thus, dominated by the much more specific verbal narrative, come to mean, that is, visualize what is *said*.

We noted previously that research on visual environmental communication has shown an increasing tendency toward abstraction or decontextualization of images *from* specific identifiable geographic or cultural environments *to* generic, iconic or "representative" global environments. This, of course, means that images become more open to multiple interpretations – polysemic – and in turn therefore, more dependent on their meaning being "anchored" (Barthes, 1977) in the accompanying text. The immediate anchoring or directing of the "reading" of the image is done through the caption, but often much more extensively through the text, particularly where the image is made to stand, symbolically, for what may be highly complex and largely invisible environmental/scientific processes and phenomena.

As we have argued elsewhere (Hansen, 2010), the visual vocabulary of climate change and other environmental issues is not one that offers itself for ready recognition or with ready-made meanings, but rather one that has to be "constructed" and where images are made to signify "climate change," "ecological threat," "contamination," "risk to health," etc. This is a process where we, as consumers of images, learn over time to associate particular images with particular environmental issues, problems or phenomena, and where at some point when these images are so much part of the recognizable and recognized public vocabulary that they need no further explanation, we can refer to them as "icons" or "symbols." According to Difrancesco and Young's (2011) analysis that point has not yet been reached in the visualization of climate change: "The issue is still awaiting its iconic image(s) with the power to steer arguments and 'speak for themselves' in the public sphere" (p. 533).

While the interaction of text/language and image/visuals has received considerable attention in the literature on visual environmental communication, there has been much less discussion of how audio/sound and color (the chromatic sign system), anchor or complement the meanings of visual representations of the environment. Yet, sound and color are clearly important to how meaning is created in environmental imagery and deserve, we would argue, a great deal more attention than they have received so far in visual environmental analysis.

Composition/Perspective/Point of View (Angle)/Gaze and Narrative

When communication researchers in the past couple of decades have defined the increasingly popular concept of "framing" in media and communication research, they often note the term's origin in visual analysis and as a reference to the frame delimiting and surrounding a photograph, painting or other visual representation. And this is of course, one of the key features of photographs: that they seemingly offer an unadulterated view/image of that which they depict, while at the same time clearly offering only a highly selective view (including and center-staging some parts, while excluding others, that is, those that lie outside the frame). But while the selectiveness of the frame is important to interpretation, the composition, point of view and perspective offered within the visual are equally significant elements (Kress & van Leeuwen, 1996/2006).

The way that photographs position the viewer in relation to that which is depicted is significant in similar ways in terms of constructing the viewer's engagement: long-distance/high-angle views construct both an impersonal relationship with what is depicted and potentially a sense of empowerment (Peeples, 2011); low-angle/upward-looking perspectives and macro close-ups, by contrast, may be used to engender a sense of optimism, intimacy and familiarity, and perhaps control through enhanced knowledge (echoing the metaphor "to take a closer look" at something in order to gain a better understanding and command of it). Both high-angle and/or aerial shots and low-angle and macro-shots also, of course are enticing and engaging in the sense that they provide the viewer with a new and unusual perspective, a new way of seeing the object, which is not normally or usually readily accessible to the viewer (Elliott, 2006).

The significance of visual analysis attending closely to the composition and point of view is that these are deliberately constructed by the producers of images with a view to achieving the desired ideological alignment and response in viewers.

Attending to narrative – the syntagmatic dimension of sign-systems – is important in visual analysis at two levels: (1) the immediate level of the story being told by individual images, or – more often – a selection or series of images; and (2) the meta-narratives that are activated by or drawn upon by the individual images analyzed. This latter dimension refers to what is also often labeled – following Gamson and Modigliani (1989) – as cultural packages (see our discussion below), discourses which summarize and encapsulate widely known and recognized ways of looking at and understanding our environment (see Ferreira, 2004, for an insightful study pressing into service narrative visual analysis and the concept of cultural resonance in the analysis of the Hollywood environmental film *Erin Brockovich*).

Contexts and Sites in Visual Analysis

Our reading of the literature on visual environmental communication leads us to conclude that, as well as attending to the immediate semiotic characteristics of visuals, visual analysis needs to focus on the contexts and sites of visual communication. Here, we delineate what we identify as the three major contexts (communicative, cultural and historical) and the three major sites (production, content, consumption) of visual meaning-making. We see these as useful for analyzing and understanding the way that visual communication works, as well as for conceptualizing and understanding the wider roles of visual communication in relation to the social and political construction of environmental issues and controversies in the public sphere.

The Communicative Context

For their meaning and "effect," to a large extent, images depend on the communicative situation or context of their communication. Conventions of genre (advertisement, news-report, art-exhibition, campaign, photographic essay, etc.) and medium (magazine, newspaper, television, website, etc.) impinge on what can be communicated visually and how. More importantly, perhaps, they set boundaries for and guide the way in which we, as viewers/consumers, make sense of images. As viewers/consumers, our awareness and knowledge of the communicative situation (including the intended purpose – for example, to inform, to sell a product/place/value, to persuade, etc. – of the communication to which we are attending) circumscribe how we make sense of, use or react to what we are seeing. The recent literature on visual environmental communication has encompassed a broad variety of media and communicative contexts, including advertising (Linder, 2006; Rutherford, 2000; Svoboda, 2011), television news (Lester & Cottle, 2009), news and other magazines (Remillard, 2011; Todd, 2010), newspapers (Boholm, 1998;

DiFrancesco & Young, 2011; Seppänen & Väliverronen, 2003; Smith & Joffe, 2009), environmental campaigns (Doyle, 2007; Manzo, 2010), and films (Starosielski, 2011). The contributions in this issue extend these developments by addressing a range of important communicative contexts and media, including environmental activism campaigns (Schwarz; Cozen), sports advertising (Ferrari), news magazines (Meisner and Takahashi), news and popular media (Peeples), and government-agency sponsored promotional campaigns (Takach; Porter).

The Cultural Context

Visual environmental depictions derive meaning and power from the extent to which they resonate (Gamson & Modigliani, 1989) with recognizable and deep-seated cultural conventions, narratives and values, and they serve to perpetuate these to the extent that they normalize, naturalize, or leave unquestioned their fundamental assumptions and world-views. Gamson and Modigliani (1989), in their pioneering study of public representations and opinion on nuclear power, show how the representation and interpretation of nuclear power are articulated through a series of cultural packages – clusters or schema of meaning, assumptions, values that are used for making sense of the world around us – including the progress package, the energy independence package, the devil's bargain package, the runaway package etc.

We can extend Gamson and Modigliani's notion of cultural packages to encompass also culturally (and sometimes nationally) specific ways of constructing or viewing nature and the environment, particularly in the tension, often observed between the utilitarian enlightenment view of nature as a resource to be controlled, dominated and exploited, a romantic and romanticized view of nature as sublime, pristine, the home of authenticity, and an ecological/environmental view of nature as fragile and in need of protection, but at the same time a resource for sustainable recreation and use. These cultural interpretations, then, not only influence how the environment and environmental issues are visualized (by photographers, media professionals, designers and others involved in visually producing/representing the environment), but in turn of course also circumscribe – through our cultural embeddedness as consumers/viewers/audiences – how we understand/interpret/make sense of visual representations of the environment.

Attending to the cultural context in visual analysis is important, not only because it provides clues to repertoires and reservoirs of meaning that are (need to be) mobilized in order to make sense of images, but significantly because they enable us to begin to tackle questions of how ideological power is projected through the naturalization/normalization of a particular view of the environment/nature at the expense of other alternative views. Several visual analysis studies have shown, for example, how the nature/environment perspectives of indigenous people are marginalized or erased altogether (Berger, 2003; DeLuca & Demo, 2000; Remillard, 2011) and how whole continents, Africa in this example, are visually constructed through the anthropocentric and cultural lens of the (Western, developed world) tourist, obscuring rather than revealing nature, the environment or indeed the damage caused by tourism (Todd, 2010).

The Historical Context

Closely linked to and intertwined with the cultural context, history provides an important and significant context for understanding and interpreting visual environmental images, as indeed it does for the analysis and interpretation of environmental communication generally. This becomes particularly evident in longitudinal studies, which demonstrate how views and representations of the environment (and nature, more specifically) change over time (Williams, 1976/1983), as demonstrated in studies of television documentaries (Wall, 1999), films (Mitman, 1999) and advertising (Ahern et al., 2012; Howlett & Raglon, 1992; Kroma & Flora, 2003; Rutherford, 2000). In an exemplary analysis of advertisements for pesticides in agricultural magazines, spanning the half century from the 1940s to the 1990s, Kroma and Flora (2003) demonstrate – through a productive combination of linguistic and visual analysis – the changing prominence of three different discourses: during the 1940s–1960s a "science" discourse articulating the post-war faith in progress through science; during the 1970s–1980s, a "control" (of nature/the environment) discourse drawing extensively from military/combat control metaphors; through, in the 1990s, a "nature-attuned" discourse reflecting environmental sensibilities – concerns about sustainability, protection of and harmony with nature – emerging during the latter half of the twentieth century. They conclude that "changing images reflect how the agricultural industry strategically repositions itself to sustain market and corporate profit by co-opting dominant cultural themes at specific historical moments in media advertising" (Kroma & Flora, 2003, p. 21).

Similar trends, including the macro shift from a romanticized through a utilitarian/science-driven/resource-focused to a sustainability/nature-attuned view have been identified in other studies of a variety of media and genres (Howlett & Raglon, 1992; Mitman, 1999; Wall, 1999). New visualizations are *produced* in response to, engaging with and often as a direct alternative to or criticism of, previous dominant visualizations. Likewise, new visualizations are understood/interpreted by audiences – not in a cultural or historical vacuum, but – against the familiarity with and knowledge of previous dominant visual discourses. Visual images, like linguistic narratives, of course are "inter-textual" – drawing on, engaging with and indeed deriving meaning from their referencing of previous, known, narratives or discourses.

Historical context – as a focus of analysis – also helps in directing attention to the particular referencing and use of *the past* that is often found in visualization of nature, the environment, the countryside, the wilderness etc. Many studies have thus pointed to the visual construction/referencing/orchestration of the past in advertising and other media as articulating a post-industrial, anti-modern (see Williams, 1973), nostalgic, romantic yearning for authenticity, identity, social cohesiveness and tradition (see Armitage, 2003; Creighton, 1997; Negra, 2001; Rutherford, 2000).

In the present issue, two essays (Peeples and Meisner and Takahashi) demonstrate significant historical changes in visual imagery, while two others (Takach and Porter) point to the importance – in visual analysis – of contextualizing and understanding the present visualizations, the object of their analysis, against the background of preceding, earlier visualizations.

Sites of Meaning-Making: Production, Content and Consumption

Rose (2012) argues that there are "three sites at which the meanings of an image are made: the site(s) of production, the site of the image itself and the site(s) where it is seen by various audiences" (p. 19). This argument is appealing, not least because it resonates well with the traditional and much-used division of principal foci in communication research: production, content and audiences. Of the three sites, the content site has attracted ample research, while the consumption site less, and the production site hardly any at all. Here, we comment briefly on the production site and the consumption site as foci for research on visual environmental communication.

While we accept Rose's model as a useful guide to the principal sites of visual meaning-making, we would also argue for a more political-economy oriented conceptualization than that which is offered by Rose, one that directs analytical attention at the production site not only to the analysis of the producers (professionals) and technical means of image production, but also to how image production and successful visually based claims-making relate to resource access. As in environmental claims-making generally (see Beder, 2002; Davis, 2003), resource access varies hugely across producers of visual environmental communication, which can range from small and perhaps primarily local groups of environmental activists, to alliances of artists, photographers and environmental activists through to large international environmental groups, advertising and media organizations, corporations/companies/ businesses (e.g., major oil companies) to governments and agencies affiliated with (local and national) the government. While studies of visual environmental communication have hinted at the production site and resource access as important factors in analysis of the construction of environmental images, few have provided analysis of how resource access, ownership and organizational affiliation impinge on their production. This is a site that deserves considerably more research attention.

While the recent decade or so has seen a number of important studies of how audiences interpret and make sense of the visuals used in environmental communication, this is a research focus that is still far less researched than the images themselves. Much of the work that has been done has drawn on psychological theories concerned with the ability of visuals – reckoned to be significantly more powerful than text in this respect – to evoke affective responses and emotions. Seppänen and Väliverronen (2003) discuss this at length in their analysis of news photographs, as do Boholm (1998), Joffe (2008), Smith and Joffe (2009), O'Neill and Nicholson-Cole (2009) and Manzo (2010). O'Neill and Nicholson-Cole (2009) empirically test audience reactions to "fear-inducing" and other images of climate change and conclude that, while fear images may be effective in attracting attention, they do not motivate personal engagement. By contrast, they find that non-threatening imagery linking to everyday emotions and concerns "tend to be the most engaging" (p. 355). These are promising avenues for visual environmental communication research on audiences, especially when integrated with and considered in the context of traditional models of media influence and tested across different cultural contexts.

Contents of this Special Edition

Elizabeth Schwartz's paper looks at the work of the International League of Conservation Photographers, who in this case seeks to create images aimed at supporting an environmental cause in a series of shows. The paper lays out the details of how photography is used for drawing attention to environmental issues pertaining to the Chesapeake Bay watershed. Through meticulous analysis of the photographs produced, Schwartz reveals the strategies the photographers employ to help to depict power relations among the stakeholders in the watershed area. The strategies may encourage audiences to develop a regional collective identity of concerned citizens who will work together to help to protect and clean up the watershed. The analysis shows how carefully crafted images work to mobilize certain kinds of symbolism in support of environmental awareness.

Jennifer Peeples' paper investigates the way that toxins, along with many other threats to the environment, are invisible and banal and thus present particular challenges for visualization. It is how toxins come to be represented visually, connecting them to different events, issues and ideas which gives them cultural resonance and an identity or even a reality in the minds of the public. Attending to the roles of composition, visual narratives, cultural and historical context, Peeples looks at the way that visual representations of Agent Orange, contradictory in terms of its effects on Vietnamese and American citizens, have come to give meaning and cultural resonance to this particular toxin.

Geo Takach's paper analyzes an official place-branding slideshow produced by the Province of Alberta (Canada), which has an economy based significantly on producing non-renewable fossil fuels. The paper examines the slideshow in terms of romantic/extractive gazes, with an interest in the way that people are removed from nature in a way that legitimizes the subordination of land to human purposes. The paper situates Alberta's rebranding on Corbett's continuum of anthropocentric-ecocentric values; interrogates connections between that rebranding effort and invisible flows of power at work in broader, underlying societal processes like globalization, Neoliberalism and consumerism.

In the following essay, Nicole Porter also examines visualization strategies in the context of place branding, in this case the Blue Mountains region in Australia.

Through a detailed analysis of the Brand Blue Mountains branding strategy and material, Porter examines how the brand aims to define the region's identity and values through a coordinated suite of visual means and content. Exercising what Porter refers to as extraordinary control over the visual expression, place brand strategists construct a selective narrative of positive nature-based sensory experience, naturalizing and reinforcing a particular market-friendly version of place.

Mark Meisner and Bruno Takahashi study the covers of *Time* magazine from 1923 to 2011, using a combination of quantitative and qualitative content analysis to explore the visual representation of nature and environmental issues. They reveal that the presence of environmental issues and nature on the covers has increased over the decades. Furthermore, *Time* takes an advocacy position on some environmental issues, but it is a shallow one that is weakly argued through less-than-engaging imagery, and the advocacy fails to offer much by way of solutions or agency to the reader.

Matthew Ferrari analyzes the way that sports commercials and related visual culture in the context of mixed martial arts (MMA) draw on and activate wildness and nature symbolism. Drawing from a range of literatures – including on gender, the primitive and the cultural production of nature and the natural – Ferrari shows how discourses of nature, gender and the environment cross paths in and inform the highly mediatized culture of sports. Carefully unpacking the ideological uses of nature in the particular selected sports-related visual culture, he shows how nature imagery is employed to authorize notions of wildness, the primitive, and maleness, and to naturalize aggression.

Brian Cozen's paper examines the Canary Project's Green Patriot Posters campaign as activist art. He examines how campaign posters are constructed, and particularly how they draw on and deploy historically older frames, to comment on contemporary relations to the environment and environmental issues. Paying particular attention to the orientational metaphors deployed, Cozen's analysis – sensitive to the ideological function of visual and narrative practices – shows how a range of visual designs question, subvert and promote continued economic growth and an antology that "more" equals "better."

Taken together the papers in this special issue greatly extend definitions of the "visual" in visual environmental communication. And they show in exemplary fashion how attention to sign systems, composition/perspective and context (communicative, cultural and historical) enable us to address key questions about the ideological and political roles of the visual in environmental communication.

References

Adam, B. (1998). *Timescapes of modernity: The environment and invisible hazards.* London: Routledge.

Ahern, L., Bortree, D., & Smith, A. (2012, May 22). Key trends in environmental advertising across 30 years in National Geographic magazine. *Public Understanding of Science.* Retrieved from http://pus.sagepub.com/content/early/2012/05/21/0963662512444848

Armitage, K. C. (2003). Commercial Indians: Authenticity, nature and industrial capitalism in advertising at the turn of the twentieth century. *The Michigan Historical Review, 29*(2), 71–95. doi:10.2307/20174034

Barthes, R. (1973). *Mythologies.* London: Fontana.

Barthes, R. (1977). Rhetoric of the image. In S. Heath (Ed.), *Image, music, text: Essays selected and translated by Stephen Heath* (pp. 32–51). London: Fontana.

Baudrillard, J. (1988). *Selected writings.* (Mark Poster, Ed.). Cambridge: Polity Press.

Beder, S. (2002). *Global spin: The corporate assault on environmentalism.* Totnes: Green Books.

Berger, J. (1972). *Ways of seeing.* London: Penguin.

Berger, M. A. (2003). Overexposed: Whiteness and the landscape photography of Carleton Watkins. *Oxford Art Journal, 26*(1), 1–23. doi:10.1093/oxartj/26.1.1

Boholm, A. (1998). Visual images and risk messages: Commemorating Chernobyl. *Risk Decision and Policy, 3*(2), 125–143. doi:10.1080/135753098348248

Bordwell, D., & Thompson, K. (2008). *Film art: An introduction.* New York, NY: McGraw-Hill.

Bousé, D. (2000). *Wildlife films.* Philadelphia, PA: University of Pennsylvania Press.

Boykoff, M. T. (2007). Flogging a dead norm? Newspaper coverage of anthropogenic climate change in the United States and United Kingdom from 2003 to 2006. *Area, 39*(4), 470–481. doi:10.1111/j.1475-4762.2007.00769.x

Boykoff, M. T. (2008). Lost in translation? United States television news coverage of anthropogenic climate change, 1995–2004. *Climatic Change, 86*(1–2), 1–11. doi:10.1007/s10584-007-9299-3

Carvalho, A., & Burgess, J. (2005). Cultural circuits of climate change in UK broadsheet newspapers, 1985–2003. *Risk Analysis, 25*(6), 1457–1469. doi:10.1111/j.1539-6924.2005.00692.x

Cottle, S. (2000). TV news, lay voices and the visualisation of environmental risks. In S. Allan, B. Adam, & C. Carter (Eds.), *Environmental risks and the media* (pp. 29–44). London: Routledge.

Cottle, S. (2009). *Global crisis reporting: Journalism in the global age.* Milton Keynes: Open University Press.

Cox, R. (2013). *Environmental communication and the public sphere* (3rd ed). London: Sage.

Creighton, M. (1997). Consuming rural Japan: The marketing of tradition and nostalgia in the Japanese travel industry. *Ethnology, 36*(3), 239–254. doi:10.2307/3773988

Davis, A. (2003). Public relations and news sources. In S. Cottle (Ed.), *News, public relations and power* (pp. 27–42). London: Sage.

DeLuca, K. M., & Demo, A. T. (2000). Imaging nature: Watkins, Yosemite, and the birth of environmentalism. *Critical Studies in Media Communication, 17*(3), 241–260. doi:10.1080/15295030009388395

DiFrancesco, D. A., & Young, N. (2011). Seeing climate change: The visual construction of global warming in Canadian national print media. *Cultural Geographies, 18*(4), 517–536. doi:10.1177/1474474010382072

Doyle, J. (2007). Picturing the clima(c)tic: Greenpeace and the representational politics of climate change communication. *Science as Culture, 16*(2), 129–150. doi:10.1080/09505430701368938

Elkins, J. (2003). *Visual studies: A skeptical introduction.* London: Routledge.

Elliot, N. L. (2006). *Mediating nature.* London: Routledge.

Ferreira, C. (2004). Risk, transparency and cover up: Media narratives and cultural resonance. *Journal of Risk Research, 7*(2), 199–211. doi:10.1080/1366987042000171294

Foucault, M. (1972). *The archaeology of knowledge.* London: Tavistock.

Gamson, W. A., & Modigliani, A. (1989). Media discourse and public opinion on nuclear power: A constructionist approach. *American Journal of Sociology, 95*(1), 1–37. doi:10.1086/229213

Hall, S. (1977). *Representation: Cultural representations and signifying practice.* London: Sage.

Hall, S. (1973/1980). Encoding/decoding. In S. Hall, D. Hobson, A. Lowe & P. Willis (Eds.), *Culture, media, language: Working papers in cultural studies, 1972–79* (pp. 128–138). Centre for Contemporary Cultural Studies. London: Hutchinson.

Hannigan, J. A. (2006). *Environmental sociology.* London: Routledge.

Hansen, A. (2002). Discourses of nature in advertising. *Communications, 27*(4), 499–511. doi:10.1515/comm.2002.005

Hansen, A. (2006). Tampering with nature: "Nature" and the "natural" in media coverage of genetics and biotechnology. *Media, Culture & Society, 28*(6), 811–834. doi:10.1177/0163443706067026

Hansen, A. (2010). *Environment, media and communication.* London: Routledge.

Hansen, A., & Machin, D. (2008). Visually branding the environment: Climate change as a marketing opportunity. *Discourse Studies, 10*(6), 777–794. doi:10.1177/1461445608098200

Howlett, M., & Raglon, R. (1992). Constructing the environmental spectacle: Green advertisements and the greening of the corporate image. *Environmental History Review, 16*(4), 53–68. doi:10.2307/3984949

Jenks, C. (Ed.). (1995). *Visual culture.* London: Routledge.

Jhally, S. (1987). *The codes of advertising: Fetishism and the political economy of meaning in consumer society.* London: Routledge.

Jhally, S. (1991). *The codes of advertising: Fetishism and the political economy of meaning in consumer society* (New ed.). London: Routledge.

Joffe, H. (2008). The power of visual material: Persuasion, emotion and identification. *Diogenes, 55*(1), 84–93. doi:10.1177/0392192107087919

Kress, G., & van Leeuwen, T. (1996/2006). *Reading images: The grammar of visual design.* London: Routledge.

Kroma, M. M., & Flora, C. B. (2003). Greening pesticides: A historical analysis of the social construction of farm chemical advertisements. *Agriculture and Human Values, 20*(1), 21–35. doi:10.1023/A:1022408506244

Lester, L., & Cottle, S. (2009). Visualizing climate change: Television news and ecological citizenship. *International Journal of Communication, 3*, 920–936.

Linder, S. H. (2006). Cashing-in on risk claims: On the for-profit inversion of signifiers for "global Warming". *Social Semiotics, 16*(1), 103–132.

Macnaghten, P., & Urry, J. (1998). *Contested natures.* London: Sage.

Manzo, K. (2010). Imaging vulnerability: The iconography of climate change. *Area, 42*(1), 96–107. doi:10.1111/j.1475-4762.2009.00887.x

McChesney, R. W. (2004). *The problem of the media: U.S. communication politics in the 21st century.* New York, NY: Monthly Review Press.

McDonald, P., & Wasko, J. (Eds.). (2007). *The contemporary Hollywood film industry.* London: Wiley.

Mellor, F. (2009). The politics of accuracy in judging global warming films. *Environmental Communication-a Journal of Nature and Culture, 3*(2), 134–150. doi:10.1080/17524030902916574

Metz, C. (1974). *Film language: A semiotics of the cinema.* (Michael Taylor, Trans.). New York, NY: Oxford University Press.

Mitman, G. (1999). *Reel nature: America's romance with wildlife on film.* Cambridge, MA: Harvard University Press.

Monaco, J. (2009). *How to read a film: movies, media and beyond.* New York, NY: Routledge.

Mulvey, L. (1975). Visual pleasure and narrative cinema. *Screen, 16*(3), 6–18. doi:10.1093/screen/16.3.6

Negra, D. (2001). Consuming Ireland: Lucky Charms cereal, Irish spring soap and 1-800-SHAMROCK. *Cultural Studies, 15*(1), 76–97. doi:10.1093/screen/16.3.6

O'Neill, S., & Nicholson-Cole, S. (2009). Fear won't do it": Promoting positive engagement with climate change through visual and iconic representation. *Science Communication, 30*(3), 355–379. doi:10.1177/1075547008329201

Panofsky, I. (1972). *Studies in iconology: Humanistic themes in the art of the renaissance.* New York, NY: Harper & Row.

Pauwels, L. (2012). An integrated conceptual framework for visual social research. In E. Margolis & L. Pauwels (Eds.), *SAGE handbook of visual research methods* (pp. 3–23). London: Sage.

Peeples, J. (2011). Toxic sublime: Imaging contaminated landscapes. *Environmental Communication-a Journal of Nature and Culture, 5*(4), 373–392. doi:10.1080/17524032.2011.616516

Remillard, C. (2011). Picturing environmental risk: The Canadian oil sands and the National Geographic. *International Communication Gazette, 73*(1–2), 127–143. doi:10.1177/1748048510386745

Rose, G. (2012). *Visual methodologies: An introduction to the interpretation of visual materials.* London: Sage.

Rust, S., Monani, S., & Cubitt, S. (2013). *Ecocinema theory and practice.* London: Routledge.

Rutherford, P. (2000). *Endless propaganda: The advertising of public goods.* Toronto, ON: University of Toronto Press.

Schoenfeld, A. C., Meier, R. F., & Griffin, R. J. (1979). Constructing a social problem – The press and the environment. *Social Problems, 27*(1), 38–61. doi:10.2307/800015

Seppänen, J., & Väliverronen, E. (2003). Visualising biodiversity: The role of photographs in environmental discourse. *Science as Culture, 12*(1), 59–85. doi:10.1080/0950543032000062263

Shanahan, J., & McComas, K. (1999). *Nature stories: Depictions of the environment and their effects.* Cresskill, NJ: Hampton Press.

Smith, N. W., & Joffe, H. (2009). Climate change in the British press: The role of the visual. *Safety, Reliability and Risk Analysis: Theory, Methods and Applications, 1–4*, 1293–1300. Retrieved form: http://dx.doi.org/10.1177/0163443706067026.

Sontag, S. (2004). *Regarding the pain of others.* London: Penguin.

Soper, K. (1995). *What is Nature?* Oxford: Blackwell.

Starosielski, N. (2011). "Movements that are drawn": A history of environmental animation from the Lorax to FernGully to Avatar. *International Communication Gazette, 73*(1–2), 145–163. doi:10.1177/1748048510386746

Svoboda, M. (2011). *Advertising climate change: A study of green ads, 2005 – 2010.* Retrieved from http://www.yaleclimatemediaforum.org/2011/07/advertising-climate-change-a-study-of-green-ads-2005-%e2%80%93-2010/

Szerszynski, B., Urry, J., & Myers, G. (2000). Mediating global citizenship. In J. Smith (Ed.), *The daily globe: Environmental change, the public and the media* (pp. 97–114). London: Earthscan Publications.

Todd, A. M. (2010). Anthropocentric distance in *National Geographic's* environmental aesthetic. *Environmental Communication: A Journal of Nature and Culture, 4*(2), 206–224. doi:10.1080/17524030903522371

Urry, J. (1992). The tourist gaze and the 'environment'. *Theory, Culture & Society, 9*(3), 1–26. doi:10.1177/026327692009003001

Wall, G. (1999). Science, nature, and the nature of things: An instance of Canadian environmental discourse, 1960–1994. *Canadian Journal of Sociology-Cahiers Canadiens De Sociologie, 24*(1), 53–85. doi:10.2307/3341478

Williams, R. (1973). *The country and the city.* London: Chatto & Windus.

Williams, R. (1976/1983). *Keywords: A vocabulary of culture and society.* London: Flamingo/Fontana.

Williamson, J. (1978). *Decoding advertisements: Ideology and meaning in advertising.* London: Marion Boyars.

Visualizing the Chesapeake Bay Watershed Debate

Elizabeth Anne Gervais Schwarz

Environmental organizations often use visual material to inform society about environmental concerns and their associated policy issues. This case study examines the process by which the International League of Conservation Photographers (iLCP) and the Chesapeake Bay Foundation (CBF) use a Rapid Assessment Visual Expedition (RAVE) to draw attention to the environmental issues surrounding the Chesapeake Bay watershed. In addition, the study analyzes the resulting photographs captured during the event. The CBF and the iLCP strategically use the RAVE to create scientific and local knowledge that they use to present their understanding of the Chesapeake Bay. An analysis of the slideshows generated from the RAVE shows how the strategies the photographers employ help to depict power relations among the stakeholders in the watershed area. The strategies may encourage audiences to develop a regional collective identity of concerned citizens who will work together to help to protect and clean the watershed.

In the summer of 2010, eight photographers from the International League of Conservation Photographers (iLCP) descended upon the Chesapeake Bay watershed to document the environmental issues surrounding the watershed. Photographers frequently focus on sociologically related topics and are often interested in contemporary social problems such as those that involve scientific, social, and political concerns (Becker, 1986; Goffman, 1979). Increasingly, environmental policy decision-making is influenced by how the public perceives related environmental issues (Hansen, 2011). Society's perceptions of nature are socially, politically, and culturally constructed, and visual artifacts play a role in these processes (Hansen & Machin, 2008). In this paper, I will analyze how the iLCP

Elizabeth Anne Gervais Schwarz is a Graduate Student at University of California Riverside.

and Chesapeake Bay Foundation (CBF) construct conservation issues through their use of a Rapid Assessment Visual Expedition (RAVE) in support of the Chesapeake Bay Clean Water and Ecosystem Restoration Act. The RAVE can be seen as a focusing event where narratives and photographs combine to draw attention to conservation problems, create a specific way of seeing issues, and motivate various audiences to act.

Recently, there has been a call for research that helps understand how environmental narratives are realized visually (Hansen, 2011; Hansen & Machin, 2008). Similarly, social movement scholars call for research examining how social movements make use of visual materials (Doerr, 2010; Philipps, 2011). Scholars recognize that environmental groups play important roles in framing environmental issues and presenting environmental narratives. Research focused on media and environmental narratives is quite prevalent. Research on the role of other players in creating environmental knowledge, such as conservation groups and environmental NGOs, is less common and offers the opportunity for future examination (Doyle, 2007; Eden, Donaldson, & Walker, 2006; Meusburger et al., 2010). This analysis will answer these calls for research.

To perform this analysis I will engage the iLCP's Chesapeake Bay RAVE, sponsored by the CBF. While there is only one actor in the Chesapeake Bay watershed policy debate, the CBF has been involved in the discussion for over 40 years and is the largest and most vocal public interest environmental group working on behalf of the Chesapeake Bay (Ernst, 2003). Therefore, the information presented by the iLCP and the CBF, and the resulting perspective they provide, represents an important viewpoint.

The Chesapeake Bay is the nation's largest estuary (Ernst, 2003). The watershed includes portions of Delaware, Maryland, New York, Pennsylvania, Virginia, West Virginia, and the District of Columbia. In total, the Chesapeake Bay watershed encompasses 64,000 square miles of land and 100,000 miles of streams and rivers (Burke & Dunn, 2010). Water quality issues have plagued the Chesapeake Bay watershed for decades and spurred the introduction of restoration projects. The Chesapeake Bay Agreement, an agreement signed in 1983 by the governors of Maryland, Virginia, Pennsylvania; the mayor of the District of Columbia; the administrator of the EPA and the chair of the Chesapeake Bay Commission, was an initial step to help improve the watershed. However, because the Chesapeake Bay covers such a large area, there are many constituencies that have interests in the watershed, including the public, agriculture, coal and steel industries, crab harvesters, and residential and commercial development groups, that often have conflicting understandings of what should happen to the watershed. In part, these differing viewpoints have made it difficult for stakeholders to develop solutions to improve the health of the watershed (Ernst, 2003).

In the following sections, I will review the role of focusing events and causal stories in agenda setting, provide more information about iLCP RAVEs, and analyze the material generated from the Chesapeake Bay RAVE.

Agenda Setting, Focusing Events, and Causal Stories

Activities in the natural world are often interpreted as "undirected, unoriented, unanimated, unguided, 'purely physical'" (Goffman, 1974, p. 22) while, alternatively, the social world is related to control and intent. When the natural world and social world come together, the result can be "unintended consequences of willed human action" (Stone, 1989, p. 285). For example, causal stories can link pollution to human behavior and not just an accident or nature (Stone, 1989). In this sense, when analyzing images as constructing an environmental issue, they must be understood within the discourse of development and sustainability as human behavior mixes with nature (Remillard, 2011). Similarly, this supports the idea of including analysis of the interaction between the physical and social worlds, or what Freudenburg, Frickel, and Gramling (1995) call conjoint constitution, when addressing environmental issues.

Drawing from agenda setting literature, Stone (1989) suggests that defining problems involves image making, a process in which images attribute cause, blame, and responsibility. The iLCP and the CBF use photographs to promote their particular interests and ideologies about the Chesapeake Bay watershed that they hope will shape their audiences' perceptions of the environment problem (Hansen & Machin, 2008). Through the use of a RAVE, the iLCP and the CBF work on agenda setting, or "the collection of activities that policy entrepreneurs engage in when they want to direct the attention of public officials...toward a particular problem" (Keller, 2009). As such, the Chesapeake Bay RAVE can be viewed as a focusing event. Focusing events are a form of agenda setting that serve as catalysts to get the attention of environmental policy-makers. They are also used as a form of evidence that demonstrates that policy needs to change (Birkland, 1998). The iLCP and CBF use the RAVE, an expedition in which photographers partner with local stakeholders to visually document an area in a short period of time and generate images that can later be used to create communications tools, as an event that can draw policy-makers' and other publics' attention to the issues surrounding the Chesapeake Bay.

While the Chesapeake Bay debate has been ongoing on for over 40 years, the introduction of the Chesapeake Bay Clean Water and Ecosystem Restoration Act made it appear that the current political will was "at the point of tipping toward long-term restoration and protection of the Bay," which made the summer of 2010 an opportune time to have this RAVE (International League of Conservation Photographers [iLCP], 2010). The iLCP and CBF then become part of a network of organizations that are involved in defining the environmental issues in the Chesapeake Bay watershed (Rabe, 2004).

A key way by which ideas enter into the policymaking process is through the development of policy narratives and causal stories. Ideally, the problem should be framed in a way that shows the negative aspects of the situation to obtain buy-in from all parties. Effective causal stories can link issues to solutions that are high on the policy agenda. The most effective stories link several perspectives, such as scientific, economic, and cultural perspectives, to the narrative (Stone, 1989).

The Role of Images

Society endows photography with certain functions, such as to inform society or represent certain aspects of society (Barthes, 1981). Photographs are thought to provide valid records of events or activities (Goffman, 1979). Barthes (1981) describes photographs as "a certificate of presence," explaining that photographs reveal for certain what has been. Photographs can also help people become aware of things they weren't previously aware of (Goffman, 1979). However, while society often accepts photographs as truthful, it is also important to consider that photographers have always had the means to manipulate their images in a variety of ways while processing their photographs. With the prevalence of photo manipulation software, such as Photoshop, it is even more important for people to think more critically about how they understand photographs (Blewitt, 2010). In addition, the final photographs and the images' messages are the product of a network of individuals, not simply the person who operates the camera (Barthes, 1977; Becker, 2008). Becker (2008) asserts that, in part, the meaning of photographs come from the organizations they are used in and the actions of all of the individuals involved in creating the image, which in this case would the CBF, iLCP, and the other organizations they work with during the RAVE.

Social movements often use visual artifacts, such as photographs, as part of their meaning making processes (Halfmann & Young, 2010; Jasper & Poulsen, 1995; O'Neill & Nicholson-Cole, 2009). During the RAVE, the photographers generated countless photographs that they used to create their understanding of the environmental issues facing the Chesapeake Bay watershed area. The iLCP and the CBF each selected specific images from the RAVE to place on their websites that help create particular policy narratives.

Ultimately, the organizations develop causal stories that provide viewers with understandings of "the way things are" (Stone, 1989). The meanings of images are fluid (Hansen, 2010) and visual objects are thought to reproduce a particular way of seeing in which images' perspectives are normalized and create a particular reality (DeLuca & Demo, 2000; Remillard, 2011). In other words, photos are based on a specific evaluation of the situation and create a "new way of seeing" (DeLuca & Demo, 2000, p. 244). Visual studies researchers often examine how images "bolster and/or disrupt long-standing social meaning" (Remillard, 2011, p. 129). Images have long been associated with social change and particularly with promoting environmentalism. In part, the impact of images depends on how well they can appeal to values and incite emotions to motivate people to act in certain ways (Müller & Backhaus, 2007). Examining the RAVE photos will provide insight into how the organizations construct the environmental issue at hand as part of a policy narrative and causal story.

The Case: The iLCP and the Chesapeake Bay RAVE

The iLCP is an example of a group of photographers that have a particular emphasis on environmental problems in society. They describe themselves as "a project-driven organization" (iLCP, 2011a). The group's Facebook profile states, "Using our photographic skills we know we can cut through the invisible boundaries of illiteracy and injustice and help bring light to the darkest corners of our planet, exposing ignorance, greed and corruption as much as beauty, wonder, and the magic of the natural world" (2011).

Their mission "is to translate conservation science into compelling visual messages targeted to specific audiences. We work with leading scientists, policy makers, government leaders, and conservation groups to produce the highest-quality documentary images of both the beauty and wonder of the natural world and the challenges facing it" (iLCP, 2011a). To reach their objectives, they "focus on understanding the turn key leverage points of a particular issue and using communication tools to create tipping points, convene audiences and move the conservation needle in significant and measureable ways" (iLCP, 2011b).

Beyond simply raising awareness, the league strives to develop products that can be used as communication tools and leveraged after events are complete. One method through which iLCP obtains visual artifacts of conservation issues are through RAVEs. RAVEs are described as "visual expeditions to places where regular reporters do not go and where the stories that matter to our planet are being told" (iLCP, 2011c). The goal is to document issues as quickly and as thoroughly as possible while also disseminating information to the media with the hope of creating a tipping point in regards to the conservation problem being documented (iLCP, 2011b). The Chesapeake Bay RAVE was one such endeavor.

The Chesapeake Bay RAVE took place from 1 August to 29 August 2010 and included the area from the headwaters of the Susquehanna River in New York to the Shenandoah Valley in Virginia (iLCP, 2011c). The photographers' charge was to visually document the area on behalf of the CBF. One desired outcome of the RAVE was for the visual media display to spur news coverage about the importance of the Chesapeake Bay Clean Water and Ecosystem Act. Ultimately, the goal of the RAVE was to use the conservation photography to help the fight to pass the Chesapeake Clean Water and Ecosystem Act (iLCP, 2010).

The CBF has spent decades trying to clean up the Chesapeake Bay and supports this remedy. The proposed Chesapeake Bay Clean Water Act would put into law many of the remedies that are already occurring in the watershed area, but more funds would go toward the cleanup. In addition, a nutrient-trading program would be created for farmers (Wood, 2010, December 27). Overall, the act would "ensure that all the states of the bay watershed develop and implement detailed plans to achieve pollution reduction targets for nitrogen, phosphorus and sediment by 2025 and would authorize much-needed funding to help cities, towns, and farmers comply with the law" (Gilliland, 2010).

The photographers captured images throughout the Chesapeake Bay watershed. A selection of the images is displayed on the iLCP and CBF websites, allowing an audience of internet users access to the photos. The iLCP slideshow contains 28 photographs while the CBF slideshow includes 31 with only six photographs overlapping. Often, captions accompany the photographs on the websites. Captions have the ability to amplify images' messages (Barthes, 1977) and can provide context to the photograph (Becker, 2007). In addition, before the RAVE takes place, information is sent out to media sources about the fact that a RAVE is going to occur in the surrounding area. A LexisNexis search for media coverage of the Chesapeake Bay RAVE, using the terms International League of Conservation Photographers and Chesapeake Bay, revealed 17 articles were written about the expedition associated with the Chesapeake Bay Clean Water and Ecosystem Restoration Act and the fact that there was an exhibit on Capitol Hill. Information drawn from these articles will also be used to expose how the Chesapeake Bay watershed issues are constructed by the organizations. Trevor Frost managed the Chesapeake Bay RAVE for iLCP. He contends,

> Stories are really important. Stories have been around since the dawn of humanity... All cultures can relate to stories and when you show one picture, that can certainly make a difference. But if you can put together a concrete story, a set of images, that pass on a certain way of thinking, it can have a bigger impact. Photos can literally be an eye-opening experience. This might sound sort of cliché, but it's amazing how many people don't know what's going on. They haven't seen it. (Wood, 2010, August 30)

Knowledge Creation

Scientific Knowledge

All of the individuals associated with the RAVE influence the final images photographers capture during the event and the eventual narratives that emerge from the series of images (Becker, 2008). This includes the types of knowledge that provides a foundation for the focusing event and subsequent images. The dynamics of the Chesapeake Bay RAVE provide insight into the role of science and knowledge in constructing the narrative of the Chesapeake Bay conservation issues. iLCP specifically states that one of their goals is to bring scientists and policymakers together. In the policy arena, the boundaries surrounding policy and science are often contentious (Keller, 2009). There is also controversy about what constitutes useable knowledge in policy decision-making (Ascher, Steelman, & Healy, 2010). Science is noted as one resource that actors often employ when trying to create their causal stories (Stone, 1989). Similar to photographs, scientific knowledge is co-produced. Scientists take cues from society as they develop the knowledge they present to the public about environmental issues (Keller, 2009). Social movement organizations and NGOs are often overlooked as scientific actors but are important in challenging dominant forms of knowledge production and reconstituting knowledge (Eden et al., 2006; Jamison, 2001).

The importance of portraying themselves as groups capable of generating legitimate scientific knowledge is seen in other instances also. An article about the RAVE included a description of specialists, a Lancaster County Conservation District Watershed Specialist and a Stream Buffer Specialist who accompanied the iLCP photographer during his shoot. Rutter (2010, August 21) writes that the pair "unspooled a yellow measuring tape and set up a laser-equipped level to gauge the shape of the stream channel." He includes a quote from one of the specialists, explaining that the Conservation Reserve Enhancement Program has helped farmers fence environmentally sensitive land and install livestock watering systems. As a result of this work, about 500 sites have been rehabilitated (Rutter, 2010, August 21). In addition, another article highlights the fact that one of the photographers "holds a doctorate and works as a marine biologist in California" (Wood, 2010, August 30).

These examples clarify the role that science plays in the science narrative that constructs environmental knowledge surrounding the Chesapeake Bay's issues. Science is seen as central to understanding what the causes and solutions to the watershed issues are. In particular, the specialist was seen as an integral part of the team and viewed as an expert about solutions to the problems facing the watershed.

Becker (1986) notes that one way photographers can improve their photographs and provide a deeper understanding of the activities they are photographing is to learn more about the subject or topic. The choices the iLCP makes in regards to which NGOs, scientists, and local experts they partner with could provide them with particular information about the environment and ultimately influence the photographs and knowledge that come from the RAVE. Focusing their careers on conservation photography could be another way iLCP photographers continue to educate themselves about the particular subjects they photograph. In addition to having a focus on conservation photography, staff members at the iLCP explain that many of the photographers have conservation backgrounds. Professional and educational backgrounds include academic credentials and experience in areas such as biology, zoology, environmental studies, and conservation advocacy. Their professional and educational backgrounds provide these iLCP photographers with unique perspectives and knowledge bases that other photographers may not have, which may influence the photographs and story generated from the RAVE (Farnsworth, 2012).

An image (Figure 1) found in the iLCP and CBF website slideshows demonstrates how science makes its way into the narratives. The image on the iLCP website, labeled "Algal Bloom in the Chesapeake Bay," is accompanied by a caption explaining the algal bloom process:

> Algal blooms flourish in the lower Chesapeake Bay following a storm that dumped nearly five inches of rain in some areas of Virginia Beach and Norfolk, Virginia, in July 2010. Here, rain, temperatures soaring above 100 degrees Fahrenheit, and pollution made for ideal algal bloom conditions, Saturday, July 31, 2010. Nicknamed the Mahogany Tides, this algae can lead to toxic conditions if consumed, raising the risk of illness and infection, as well as blocking light needed by other marine creatures. As the algae dies, the organisms will also deplete oxygen levels, putting other wildlife at risk. (iLCP, 2010)

The caption helps contextualize what the photographers captured in the photograph. Simply looking at the image, audiences with an untrained eye may recognize the economic story being told and notice the murky water that takes up the majority of the image, but miss the scientific story that the caption expands upon. The caption

Figure 1. (Color online) Algal Bloom in the Chesapeake Bay; Photographer: Morgan Heim/iLCP.

provides more detail to the story as it describes what process led to the algal blooms in the photograph and explains the possible consequences of the algal blooms. The background details that the caption provides about the process by which algal blooms are produced are vital to understanding the full story the iLCP is trying to convey.

Local Knowledge

It is important to point out that the iLCP RAVE specifically created local knowledge, which is characterized as place-based or contextual. The inclusion of local knowledge is often thought to lead to more sustainable environmental practices. In part, partnering with the CBF gives the iLCP RAVE the opportunity to develop local knowledge, but because the RAVE was funded by the CBF, bias may result in the particular knowledge created and presented (Ascher et al., 2010). According to staff at the iLCP, for this RAVE, photographers used their own photographic equipment, including camera bodies, lenses, filters, and underwater housing. The CBF provided access to boats that photographers used to navigate throughout the watershed area when necessary. To help photographers obtain the aerial photographs, the volunteer-based environmental aviation organization LightHawk donated a plane and pilot during the RAVE.

Through their use of photography, iLCP photographers create place-based knowledge that can be used in policy development discussions. As previously noted, the partnerships with local or scientific individuals influence the final images generated from the RAVE. Articles covering the RAVE mention local organizations iLCP teamed up with throughout the RAVE, such as the Lancaster County Conservation District and the Eastern Pennsylvania Coalition for Abandoned Mine Reclamation (Rutter, 2010, August 24; Legere, 2010, August 14). Local impact images are regarded as vital to help create local relevance (O'Neill & Nicholson-Cole, 2009). According to staff at the iLCP, the local organizations provide logistics and their own local conservation perspective. In the case of the Chesapeake Bay RAVE the CBF was the major player in this role. The CBF also helped to identify themes of what they thought would be important to cover during the RAVE. The fact that the Chesapeake Bay RAVE only had one major partner allowed for increased creative opportunities for the photographers.

In creating local knowledge, the iLCP's Chesapeake Bay RAVE highlights some of the spatial challenges associated with the Chesapeake Clean Water and Ecosystem Restoration Act. The Chesapeake Bay issues are not constrained by geographical boundaries, which mean they also transcend political boundaries (Rabe, 2004). The eight photographers spread out across the six states affected by the watershed issues. While the issue is contained within a single country, the photos reveal that not every state is affected the same way and various types of interests are found across the affected states. Images represent groups such as the public, farmers, crab harvesters, and the coal industry that all have a stake in the future of the Chesapeake Bay watershed. Miguel Angel de la Cueva, an iLCP photographer from Mexico explains, "I think when it comes to conservation, there's no international boundaries" (Rutter, 2010, August 21).

Constructing the Watershed: Problems, What's at Stake, Causes, and Solutions

Scholars in the domain of discourse analysis have expanded their work to examine visual material as they would linguistic text (Kress & van Leeuwen, 2006). An analysis of the photos selected from the RAVE and placed in slideshows on CBF's and iLCP's websites provide additional insight into the causal stories they created surrounding the watershed (Chesapeake Bay Foundation [CBF], 2011; iLCP, 2010). In particular, they reveal how they see the watershed, what they see as influencing the polluted watershed, and the harm being done. Furthermore, the images provide insight into the ways the physical and social worlds interact (Freudenburg et al., 1995).

The Problem

According to the photographs in the CBF slideshow, the rise in sea level leading to wetland and residential flooding, algal blooms, silt, and urban run-off are issues facing the watershed. Five out of the 31 photographs in their slideshow identify what is wrong with the watershed. For example, one image of a Maryland wildlife refuge includes a caption that explains, "Rising sea level is drowning what were vast expanses of wetlands in the Blackwater National Wildlife Refuge in Cambridge, Maryland" (CBF, 2011).

Alternatively, the iLCP's visual narrative presents water pollution from mining and algal blooms as issues facing the watershed. Four out of the 28 photographs in their slideshow address what is currently wrong with the watershed. While the audience could most likely look at the image of urban run-off in a river and identify it as a problem, with other images, such as the wildlife refuge photograph described above, the caption is needed to reposition the otherwise aesthetically pleasing landscape image as problematic.

Social movements have often used grotesque or shocking images (Halfmann & Young, 2010; Jasper & Poulsen, 1995). The images the groups offer are not necessarily overly shocking or grotesque. However, although images showing disturbing items, like orange debris from mining, may initially be a turn-off for audiences, showing the effects of various sources of pollution in a narrative format, or placing the disturbing in a familiar form (Polletta, 1998) may lead the audience to agree with the organizations' agenda.

Causes

The organizations also illustrate what they see as the influences of the contamination of the watershed in the slideshows. All of these images serve as reminders to people about how many different interests are involved in the Chesapeake Bay watershed. The CBF allocates five photographs to identifying the causes of the problems in the watershed. They point to industry (natural gas drilling and steel plants), residential development, and cattle as culprits. The iLCP includes five images that point to the causes of the environmental issues. They highlight the mining industry, residential development, and vehicle pollution as problematic.

CBF describes the influence of the Bethlehem Steel Plant (Figure 2): "The Bethlehem Steel plant has been a major economic driver as well as pollution source

Figure 2. (Colour online) Bethlehem Steel plant; Photographer: Garth Lenz/iLCP.

on the banks of the Patapsco River in Baltimore, Maryland" (CBF, 2011). Another photo depicts a natural gas drilling rig situated between farms and forests in rural Pennsylvania. The CBF also identifies aspects of farming as causes of the pollution. A photo of cows standing in a stream includes the caption: "Cattle can pollute streams with manure and cause stream bank erosion, which results in nitrogen and silt flowing down rivers and into the Chesapeake Bay" (CBF, 2011). Development is portrayed as a problem as well. Under a photo that shows an aerial view of a residential development is the text: "Sprawl development encroaches on productive Pennsylvania farmland" (CBF, 2011). Showing the effect of storm runoff, they also include an image of garbage from storm runoff in the Hickey Branch of the Anacostia River in Washington D.C. They note this garbage may end up in the Chesapeake Bay.

The images iLCP chose to incorporate into the photo slideshow on their website to portray influences of contamination are slightly different (iLCP, 2010). The influence of the coal industry is depicted in one image entitled "Coal pollution on the Elizabeth River" along with the following text:

> Norfolk Southern's Lamberts Point Coal Pier 6 Coal dust coats the Elizabeth River, runoff from the Lamberts Point's Pier 6 in Norfolk, Virginia. In September of 2005, this pier loaded its billionth ton of coal making it the largest transloading coal pier in the northern hemisphere, according to a PRNewswire story. Coal arriving at this pier comes from mining operations in Virginia, West Virginia and Kentucky primarily, and is destined for ports in five continents. This facility is capable of loading 8,000 tons of coal per hour. The amount of coal that has passed through this pier since it opened could form a train 104,000 miles long, according to the story, enough to circle the Earth four times. (iLCP, 2010)

The majority of the photographs showing industrial environments are shot from a distance and at a high angle. The long shot suggests an impersonal relationship between the audience and industry and creates a kind of visible barrier between the audience and what is depicted in the photograph. The high angle suggests that the audience has power over industry, at least symbolically (Kress & van Leeuwen, 2006). Based on this understanding, the environmental organizations may be trying to encourage a collective identity wherein the audience comes together against the industries that are causing problems in the watershed.

iLCP also identifies development and people in the community as offenders. They include an aerial photo (Figure 3) of "Marinas and McMansions." The caption beside the photo explains, "The shear density of marinas and high-end apartments built right up to the water's edge are sources of water pollution in Norfolk, Virginia" (iLCP, 2010). They also include a picture titled "Runoff from Above" looking up at a bridge running over the Elizabeth River. The accompanying text describes the scene:

> Traffic cruises over the Elizabeth River in Norfolk, Virginia. Dirt, oil and other emissions can filter down from the grating directly into the river, but that's not the only way automobiles contribute to pollution. Every vehicle on every road contributes to runoff of pollution into the Chesapeake Bay's waterways, and often traffic along the Lower Chesapeake includes bumper-to-bumper congestion. (iLCP, 2010)

Figure 3. (Colour online) Marinas and McMansions; Photographer: Morgan Heim/iLCP.

These two images of development and people in the community show aspects of society that people are used to seeing in their daily lives, but they are presented from unique perspectives. People may be aware of the marinas and residential development along the water but through the use of the aerial photograph the audience gets to see the sheer number of boats and cookie cutter residences in a pattern that one could imagine going on indefinitely. This new viewpoint could cause audiences to think about development as a negative influence on the bay's health and the high angle shot may indicate the photographers hope audiences feel they have power over development (Kress & van Leeuwen, 2006). Similarly, while residents may be used to driving over or sitting in traffic on the bridge crossing the Elizabeth River, they most likely have not had the opportunity to look up from below the bridge, from the viewpoint of being in the river. This may lead them to think differently about the relationship between the traffic, runoff, and the river below. However, this photograph is shot from a low angle, which suggests the audience may be disempowered, or those who typically have that perspective under the bridge may not have the power to influence the people in the traffic above (Kress & van Leeuwen, 2006).

Although residents may not previously have thought of development, marinas, and traffic as offenders, they may think more critically about the role these aspects of society play in the health of the watershed after seeing the images. The different viewpoints employed in these images may be enough to jar individuals out of how they would typically view various aspects of their lives and see them in a new way that aligns with the message of the environmental organizations.

What's at Stake?

The majority of the photographs in both slideshows depict what might be lost if the watershed continues to deteriorate. Twenty-one of the 31 photographs in the CBF slideshow address what's at stake. Eighteen of the 28 photographs on the iLCP slideshow address what might be lost. These fall into three general categories: nature, livelihood, and recreation.

Photos sustain the pristine, sublime depictions of nature that have been shared throughout history (DeLuca & Demo, 2000; Hansen & Machin, 2008; Linder, 2006; Doyle, 2007; Remillard, 2011). A photo of a pristine landscape on the iLCP slideshow illustrates the idyllic hills surrounding the Chesapeake Bay watershed. Another photo (Figure 4) found in both of the slideshows depicts wetlands at the mouth of the Nansemond River. The text accompanying the photo in the iLCP slideshow reveals:

> Pristine wetlands frame the mouth of the Nansemond River, a tributary of the James River near Suffolk, Virginia. This river is one of the most untouched areas in the Chesapeake Bay and a haven for wildlife and nature lovers alike. It is also the river where John Smith experienced some of his most famous encounters with Chief Powhatan. (iLCP, 2010)

Through the use of images and captions, the iLCP promotes an understanding of the Chesapeake Bay watershed as unspoiled by human interaction. Wide, sweeping,

Figure 4. (Colour online) Nansemond River; Photographer: Morgan Heim/iLCP.

aesthetically pleasing, landscape photographs paint a picture of nature as sublime. Drawing upon historical figures, such as John Smith and Chief Powhatan, also generates feelings of nostalgia. Looking at the images and reading the captions, the audience may be moved to want to keep these pastoral areas untouched, or restore damaged areas of the watershed area to their former healthy condition.

Also focused on nature, photographs reveal the plants and animals that make the watershed their home. The CBF nature images are split fairly equally between landscape and plants and animals. They include images of fish, birds, and trees. Most of the iLCP nature photographs focus on non-landscape photos. These include images of fish, birds, snails, crabs, and plants. The photographers again make use of angles and different viewpoints to engage the audience with these images of plants and animals. In opposition to earlier photographs that gave power to the audience over industrialization, in the majority of cases the plants and animals are photographed from a low angle or straight on, suggesting that they have power or are at least as important as the audience and should be seen as important (Kress & van Leeuwen, 2006).

Drawing from the iLCP slideshow, one photograph of a lotus poking through the water offers a view from the level of the plant. Another image of a crab scurrying across the sand places the audience at the level of the crab. The caption explains:

> A ghost crab makes a break for it under torchlight, heading for the sirens call of the surf at Bay's edge. Dusk heralds the time for ghost crabs to emerge from their sandy burrows and search the beaches for food, mates, and to lay their eggs. Take a walk along First Landing State Park in Virginia Beach, Virginia, and you're likely to see one after the other skittering away under beam of a headlamp.

These photographic strategies may draw audiences to the perspective of the plants and animals that are found in the watershed area. Typically, audiences would not see these aspects of the Chesapeake Bay from these unique perspectives. Seeing the world from a low angle, a different perspective than they might typically, may lead the audience to identify with organizations' narratives and see the plants and animals as valuable.

The Chesapeake Bay watershed is also constructed as a natural resource. Natural resources are considered socially and physically constructed features of the environment that are perceived to have economic value (Freudenburg et al., 1995). Images depict the watershed as an environment that encompasses the activity of humans as well as other species. The issue is constructed as one that needs to balance ecological needs and human needs (Feldman, 2011, May). The slideshows include images that depict recreation and livelihood that could be lost unless action is taken.

Images on the CBF slideshow show five different recreation photographs, including a young boy fishing, boys playing on a beach, and people kayaking. Similarly, the iLCP includes three recreation photographs that portray fishing and children playing in the water. These photographs show the positive relationships people can have with the Bay, showing appealing scenes such as children laughing in the water and serene views of people fishing. Audiences may view these photographs and place themselves in the same situations.

The CBF slideshow highlights crab harvesters, farmers, and fishermen (2011). Images on the iLCP's slideshow depict farmers and crab harvesters (iLCP, 2010). In particular, the idea of crab harvesting is deeply entrenched in the general culture of the Chesapeake Bay (Ernst, 2003). The photographers are drawing on the assumption that audiences will identify with this familiar cultural and economic icon and identify improving the Bay as a worthwhile cause.

Kress and van Leeuwen (2006) distinguish between demand images, in which the gaze is at the viewer, and offer images, where there is an absence of gaze at the viewer. Demand images acknowledge the audience explicitly and is said to do something to the audience. Alternatively, no contact is made with the audience in offer images. Instead, the item represented in the image is seen as pieces of information or something to think about. All of the photographs that include people are offer images, suggesting the photographers are providing these photographs as information about what might be lost if action is not taken. Individuals are asked to draw their own conclusions.

Depicting what might be lost may be seen as trying to appeal to values and emotions. Public perception of environmental issues has gained prominence in policy debates and "winning hearts and minds" (Hansen, 2011, p. 8) of the public may be equally as important as offering a scientific story. Visual environmental narratives may help facilitate this process. In part, the impact of images depends on how well they can appeal to values and incite emotions to motivate people to act in certain ways (Müller & Backhaus, 2007). The focus on photographs reminding audiences what is at stake provides an example of this.

Solutions

One of the main differences between the photo slideshows on the iLCP and CBF websites is that images and text from iLCP's website provide information about what is currently being done to help the watershed. The images iLCP include in their slideshow could also be seen as empowering people and showing that the problem can be remedied (Stone, 1989). Scholars tend to distinguish between fear and challenge messages (Tomaka, Blascovich, Kelsey, & Leitten, 1993). Challenge messages incorporate fear along with achievable solutions. Individuals feel they have the resources to overcome the risk the message communicates which can be highly motivating. More recent research suggests it is important to balance overly negative or fearful representations which can disempower or overwhelm people, with more empowering depictions of the situation. Fear-evoking images may get peoples attention at first but they do not have as much potential for creating action (O'Neill & Nicholson-Cole, 2009). Often, images that show solutions to environmental issues evoke feelings of personal efficacy (O'Neill & Nicholson-Cole, 2009). All of the photographs that show solutions are photographed from straight on, suggesting they are part of the reality in which audiences currently reside. This may help audiences believe that there are effective solutions to the problems found in the watershed area and feel that it is worthwhile to take action.

For example, one image titled "Old Oak in a Subdivision," depicts an East Beach housing development (Figure 5) in Norfolk, Virginia. The copy describes that the development incorporates storm runoff mitigation into its building plans:

Figure 5. Old Oak in a Subdivision; Photographer: Morgan Heim/iLCP.

Streets slope inward to help direct water flow; homes are built to maximize density while leaving room for more green space, which helps filter storm runoff before it reaches waterways. (iLCP, 2010)

Another image of the Nixon family of crab harvesters entitled "Crabbing family of the Chesapeake" includes a caption that explains how blue crab populations in the bay are coming back because of restoration efforts but that more will need to be done to ensure sustained growth (iLCP, 2010). Both of these images provide audiences with the understanding that there is hope for the watershed. The picture of the housing development, an easily recognizable image of residents around the watershed, suggests that there may be solutions in which people and the environment can coexist. Blue crabs are a familiar cultural and economic symbol of the Chesapeake Bay (Ernst, 2003). The photograph and caption describing the success of the blue crab provides an understanding that the restoration being done is working, but it must continue for the symbolic blue crab to survive.

However, the photo presentation on the CBF website only contains one image that explicitly shows any type of success. An aerial shot of a sewage treatment facility contains the caption, "A Baltimore County, Maryland, sewage treatment facility reduces nitrogen from wastewater" (CBF, 2011). Generally, effective science narratives contain three elements: causal stories, harmful effects, and how to mitigate the problem (Keller, 2009). While the images on both websites draw attention to the causes and harmful effects of the Chesapeake Bay watershed conservation issues, the CBF provides less of a focus on mitigation of the problem. The iLCP presentation begins to address mitigation through an overview of what has been successful. The absence of focusing on mitigation in the presentation of the photos may be because, as stated in the overview of the RAVE on the iLCP's website and in various articles about the RAVE, the CBF sees the solution as the passing of the Chesapeake Bay Clean Water and Ecosystem Restoration Act (iLCP 2010; Rutter, 2010, August 21; Tangley, 2011, May 1; Wood, 2010, August 30; Wood, 2010, December 27).

The Outcome

In addition to the photo slideshows that the groups placed on their websites, an exhibit of approximately 30 photographs from the RAVE was displayed from 20 September to 1 October in the Russell Rotunda on Capitol Hill, Washington D.C. Seventeen articles were written mentioning the RAVE during and after the RAVE occurred, covering four states and the District of Columbia.[1] Ultimately, the Chesapeake Clean Water and Ecosystem Act was passed over as congress people and senators broke for holiday recess. Instead, the Environmental Protection Agency (EPA) proposed a pollution diet for the states (Wood, 2010, 27 December).

Event-based tactics, such as the RAVE, are not considered the best strategy to communicate about long-term and accumulative environmental problems (Doyle, 2007). However, a spokesperson for CBF contends the images generated from the RAVE will remain important in their struggle to clean up the Chesapeake Bay watershed. They explain the images will continue to be used "in newsletters, magazines, calendars and videos and at receptions and presentations to state legislators" (Tangley, 2011, 1 May). This aligns with Stone's (1989) assertion that causal stories need to be maintained. The images will be used to sustain CBF's construction of the issues facing the Chesapeake Bay watershed.

Conclusion

The findings of this study add to existing literature in multiple ways. This case study provides insight into the process by which two environmental organizations develop environmental knowledge through a RAVE. In addition, it offers the opportunity to examine the visual narratives the organizations create using the visual material generated from the RAVE. By doing so, it adds to the literature on how environmental organizations create environmental knowledge. Next, it extends scholarship on how environmental narratives are realized visually by analyzing slideshows of photographs generated through the RAVE. Finally, this study adds to the scant research that examines how social movements use visual material as part of their meaning making processes.

The CBF and the iLCP use a RAVE to draw attention to the environmental issues facing the Chesapeake Bay watershed at a time when policy decision-making was to take place. As part of their agenda setting efforts, the groups use the RAVE as a focusing event to assert that policy needs to change (Birkland, 1998). The CBF and the iLCP use the RAVE to create scientific and local knowledge that they use to present their understanding of the Chesapeake Bay. An analysis of the slideshows generated from the RAVE shows how the photographers' strategic photographic choices help depict the power relations among the stakeholders in the watershed area and try to encourage the audience to develop a regional collective identity of concerned citizens by agreeing with the environmental narratives they present about the bay.

According to Hannigan (2006), there are six factors that interact to successfully construct an environmental issue. The RAVE facilitates many of these processes. First, there must be scientific authority for and validation of the claims associated with the issue. The groups frame and highlight their own scientific expertise and their partnerships with scientific actors. During the RAVE, the iLCP and CBF work with local organizations and scientific actors to help legitimate the environmental knowledge incorporated in their visual narratives. They appeal to the legitimacy of scientific expertise. This helps the audience may be more likely to accept the stories the iLCP and CBF present as scientific fact.

Second, there must be popularisers who can bridge environmentalism and science. Both the iLCP and CBF are well known organizations that have histories of promoting environmentalism. They each act as translators who take the scientific information necessary to understand the story of the bay and represent the issues as socially relevant in the form of a visual narrative that can be understood by a lay audience. Third, media attention should frame the issue as important. The iLCP has a goal of obtaining (positive) media attention and works vigorously to draw the media's attention on the important policy issues they are engaging with during the RAVE. In this case, the 17 articles that covered the RAVE did in fact depict the issues facing the watershed as important.

Fourth, the problem must be dramatized in highly symbolic and visual terms. The iLCP and CBF specifically use visual communication tools to tell their story of the watershed. They incorporate highly symbolic aspects, including local recreation, industry, and cultural representations. Fifth, there must be economic incentives for taking positive action. Economic factors, such as industry and livelihoods, are portrayed in the visual narratives. In particular, the narratives highlight the precarious position of the crabbing industry, which is an important industry in many regions of the watershed. And finally, an institutional sponsor must be used to legitimize and continue the process. This could be viewed as one of the goals of the RAVE and perhaps what the iLCP and CBF hope to have come out of the RAVE.

This research also provides insight into how environmental narratives are realized visually. The slideshows reveal what the problem is, who is to blame, what is at stake, and offers possible solutions. The power of images comes from their combination of fact and emotion (DiFrancesco & Young, 2011) and the RAVE photographers take on the role of scribes and poets as they document the watershed area (Sontag, 1977). While previous social movement research examined the use of grotesque images used to administer moral shock (Halfmann & Young, 2010; Jasper & Poulsen, 1995) this analysis extends the literature on social movements and visual material beyond these types of shocking images.

Instead of focusing on photograph's ability to shock, the majority of the images in both slideshows tap into audiences' hearts and minds by emphasizing what might be lost if action is not taken. This often came in forms of giving power to plants and animals that make the watershed their home by using low angled, medium to close up shots. The imagery appeals to local traditions and economic interests by drawing on common historical, cultural, and economic understandings of the watershed. This aligns with the assertion that images are useful for drawing on values and emotions that may incite action (Müller & Backhaus, 2007). Alternatively, long shots create difference between industry and audiences. By using high angles, the photographers suggest that the audience has power over industrial polluters. While recreation and industry are both ways in which the bay can be considered a natural resource, the photographers' visual strategies help audiences differentiate between them.

The iLCP's narrative provides audiences with a more explicit understanding of effective solutions that they can witness in their current lives, emphasizing the fact that all is not lost and it is worth fighting for the watershed. A combination of the use of all of these strategies may lead audiences to agree with the iLCP's story. Audiences may develop a collective identity as concerned citizens and work together to fight to help the watershed area.

Thinking about the context of the CBF can also provide greater understanding as to why the iLCP and CBF ultimately present different narratives. The CBF is considered a successful interest group, but they are not always seen as a successful political organization because "Bay-related issues involve the legislative and executive bodies of six states and the federal government" (Ernst, 2003, p. 43). The broad constituency and often divided interests that the CBF represents may constrain the type of narrative they present about the watershed as they try not to alienate certain interests found in their membership base, while the iLCP does not have the same limitations. The iLCP may be better positioned to create and advocate for alternative ways of seeing problems and solutions (Brulle, 2010), including solutions that are high on the policy agenda (Stone, 1989). Although images are said to reduce complexity in messages, there are still important complexities involved in using environmental visual narratives that need to be explored, as this case study illustrates.

Acknowledgements

A previous version of this article was presented in the Collective Behavior and Social Movement Section Roundtable: Environmentalism and Environmental Justice at the ASA Conference in Denver, CO in August 2012. The author would like to thank the International League of Conservation Photographers for their cooperation. The author would also like to thank David Feldman, Scott Brooks, Katja Guenther, and Ellen Reese for helpful feedback on earlier drafts of this paper as well as the article referees for their comments.

Note

[1] The 17 articles do not represent an increase in media coverage about the Chesapeake Bay's environmental issues. A LexisNexis search reveals a negligible decrease in coverage during the time of the RAVE (1 July–31 October 2010) compared to four months prior to the RAVE (1 March–30 June 2010).

References

Ascher, W., Steelman, T., & Healy, R. (2010). *Knowledge and environmental policy: Re-imagining the boundaries of science and politics*. Cambridge, MA: MIT Press.
Barthes, R. (1977). *Image, music, text*. New York, NY: Hill and Wang.
Barthes, R. (1981). *Camera Lucida: Reflections on photography*. New York, NY: Hill and Wang.
Becker, H. (1986). *Doing things together: Selected papers*. Evanston, IL: Northwestern University Press.
Becker, H. (2007). *Telling about society*. Chicago, IL: University of Chicago Press.
Becker, H. (2008). *Art worlds*. Los Angeles, CA: University of California Press.
Birkland, T. A. (1998). Focusing events, mobilization, and agenda setting. *Journal of Public Policy*, *18*(1), 53–74. doi:10.1017/S0143814X98000038
Blewitt, J. (2010). *Media, ecology and conservation: Using the media to protect the world's wildlife and ecosystems*. Cornwall, UK: TJ International.
Brulle, R. J. (2010). From environmental campaigns to advancing the public dialog: Environmental communication for civic engagement. *Environmental Communication*, *4*(1), 82–98. doi:10.1080/17524030903522397
Burke, D. G., & Dunn, J. E. (Eds.). (2010). *A sustainable Chesapeake: Better models for conservation*. Arlington, VA: The Conservation Fund.

Chesapeake Bay Foundation. (2011). RAVE Photo Gallery. Retrieved May 28, 2011, from http://www.cbf.org/Page.aspx?pid=2040

DeLuca, K. M., & Demo, A. T. (2000). Imaging nature: Watkins, Yosemite, and the birth of environmentalism. *Critical Studies in Media Communications, 17*(3), 241–260. doi:10.1080/15295030009388395

DiFrancesco, A. D., & Young, N. (2011). Seeing climate change: The visual construction of global warming in Canadian national print media. *Cultural Geographies, 18*(4), 517–536. doi:10.1177/1474474010382072

Doerr, N. (2010). Politicizing precarity, producing visual dialogues on migration: Transnational public spaces in social movements. *Forum: Qualitative Social Research, 11*(2), 30–57.

Doyle, J. (2007). Picturing the clima(c)tic: Greenpeace and the representational politics of climate change communication. *Science as Culture, 16*(2), 129–150. doi:10.1080/09505430701368938

Eden, S., Donaldson, A., & Walker, G. (2006). Green groups and grey areas: Scientific boundary work, nongovernmental organizations, and environmental knowledge. *Environment and Planning A, 38*(6), 1061–1076. doi:10.1068/a37287

Ernst, H. R. (2003). *Chesapeake Bay blues: Science, politics, and the struggle to save the bay.* Boulder, CO: Rowman & Littlefield.

Facebook. (2011). International League of Conservation Photographers ILCP. Retrieved from https://www.facebook.com/conservationphotography

Farnsworth, B. (2012). Conservation photography as environmental education: Focus on the pedagogues. *Environmental Education, 17*(6), 769–787.

Feldman, D. (2011, May). *International and trans-boundary environmental policy-making.* Environmental Politics and Policy Lecture conducted from the University of California, Irvine, California.

Freudenburg, W. R., Frickel, S., & Gramling, R. (1995). Beyond the nature/society divide: Learning to think about a mountain. *Sociological Forum, 10*(3), 361–392. doi:10.1007/BF02095827

Gilliland, D. (2010). Artists hope images of river live capture lawmakers' attention. *The Patriot News.* Retrieved from http://www.pennlive.com/midstate/index.ssf/2010/09/artists_hope_images_of_susqueh.html

Goffman, E. (1974). *Frame analysis.* New York, NY: Harper and Row.

Goffman, E. (1979). *Gender advertisements.* Cambridge, MA: Harvard University Press.

Halfmann, D., & Young, M. (2010). War pictures: The grotesque as a mobilizing tactic. *Mobilization: An International Quarterly, 15*(1), 1–24. Retrieved from: http://mobilization.metapress.com/openurl.asp?genre=article&issn=1086-671X&volume=15&issue=1&spage=1

Hannigan, J. (2006). *Environmental sociology.* New York, NY: Routledge.

Hansen, A. (2010). *Environment, media and communication.* New York, NY: Routledge.

Hansen, A. (2011). Communication, media, and environment: Toward reconnecting research on the production, content and social implications of environmental communication. *The International Communication Gazette, 73*(1, 2), 7–25. doi:10.1177/1748048510386739

Hansen, A., & Machin, D. (2008). Visually branding the environment: Climate change as a marketing opportunity. *Discourse Studies, 10*(6), 777–794. doi:10.1177/1461445608098200

iLCP. (2010). Chesapeake Bay RAVE. Retrieved May 28, 2011, from http://www.ilcp.com/projects/chesapeake-bay-rave

iLCP. (2011a). About. Retrieved May 28, 2011, from http://www.ilcp.com/about

iLCP. (2011b). Organization. Retrieved May 28, 2011, from http://www.ilcp.com/about/organization

iLCP. (2011c). Projects. Retrieved May 28, 2011, from http://www.ilcp.com/projects

Jamison, A. (2001). *The making of green knowledge: Environmental politics and cultural transformation.* New York, NY: Cambridge University Press.

Jasper, J. M., & Poulsen, J. D. (1995). Recruiting strangers and friends: Moral shocks and social networks in animal rights and anti-nuclear protests. *Social Problems, 42*(4), 493–512. doi:10.2307/3097043

Keller, A. C. (2009). *Science in environmental policy.* Cambridge, MA: MIT Press.

Kress, G., & van Leeuwen, T. (2006). *Reading images: The grammar of visual design*. New York, NY: Routledge.

Legere, L. (2010, August 14). Photographers work with Chesapeake Bay group to document watershed. *The Times-Tribune*. Retrieved from http://thetimes-tribune.com/photographers-work-with-chesapeake-bay-group-to-document-watershed-1.946521

Linder, S. H. (2006). Cashing-in on risk claims: On the for-profit inversion of signifiers for global warming. *Social Semiotics, 16*(1), 103–132. doi:10.1080/10350330500487927

Meusburger, P., Livingstone, D., Jöns, H., & Eden, S. (2010). NGOs, the science-lay dichotomy, and hybrid spaces of environmental knowledge. *Geographies of Science, 3*, 217–230.

Müller, U., & Backhaus, N. (2007). The Entlebuchers: People from the back of beyond? *Social Geography, 2*(1), 11–28. doi:10.5194/sg-2-11-2007

O'Neill, S., & Nicholson-Cole, S. (2009). "Fear won't do it": Promoting positive engagement with climate change through visual and iconic representations. *Science Communication, 30*(3), 355–379. doi:10.1177/1075547008329201

Philipps, A. (2011). Visual protest material as empirical data. *Visual Communication, 11*(3), 3–21. doi:10.1177/1470357211424675

Polletta, F. (1998). "It was like a fever..." narrative and identity in social protest. *Social Problems*, 137–159. doi:10.2307/3097241

Rabe, B. G. (2004). *Statehouse and greenhouse: The emerging politics of American climate change policy*. Washington, DC: Brookings Institute.

Remillard, C. (2011). Picturing environmental risk: The Canadian oil sands and the National Geographic. *The International Communication Gazette, 73*(1–2), 127–143. doi:10.1177/1748048510386745

Rutter, J. (2010, August 21). Taking their best shots. *Sunday News*. Retrieved from http://lancasteronline.com/article/local/279858_Taking-their-best-shots.html

Sontag, S. (1977). *On photography*. New York, NY: Farrar, Straus, and Giroux.

Stone, D. A. (1989). Causal stories and the formation of policy agendas. *Political Science Quarterly, 104*(2), 281–300. doi:10.2307/2151585

Tangley, L. (2011, May 1). Donating both their time and high-quality images, professional photographers are creating swat teams to help save imperiled ecosystems around the world. *State News Service* issued by the National Wildlife Federation. Retrieved from http://www.nwf.org/News-and-Magazines/National-Wildlife/PhotoZone/Archives/2011/Conservation-Photography-RAVEs.aspx

Tomaka, J., Blascovich, J., Kelsey, R. M., & Leitten, C. L. (1993). Subjective, psychological, and behavioral effects of threat and challenge appraisal. *Journal of Personality and Social Psychology, 65*(2), 248–260. doi:10.1037/0022-3514.65.2.248

Wood, P. (2010, August 30). Chesapeake Bay gets ready for a close-up. *The Capital Gazette*. Retrieved from http://www.capitalgazette.com/news/chesapeake-bay-gets-ready-for-a-close-up/article_c4fd9753-2a3a-517d-b1be-0ca7e403fd1b.html

Wood, P. (2010, December 27). Chesapeake bill dies on Capitol Hill. *The Capital Gazette*. Retrieved from http://www.capitalgazette.com/news/chesapeake-bill-dies-on-capitol-hill/article_5ed0d29a-8524-5973-a4a8-7edc37eed733.html

Imaging Toxins

Jennifer Peeples

This essay examines how toxins are visually represented in news and popular media. More specifically, it analyzes the function of visual narratives, identity, place, and uncertainty in the construction of the controversial toxicant Agent Orange, a defoliant used by the US military during the Vietnam War to reduce jungle cover and destroy cropland causing devastating health and environmental effects. Toxins present an interesting challenge for visual construction in that they are often invisible and banal in their esthetics. The essay concludes with five observations for understanding the relationship between images and toxins.

In Rachel Carson's (1962) influential "Fable for Tomorrow," she described a town stricken by a mysterious malady, one that withers crops, kills animals, and sickens humans. She shocked her readers when she explained that the devastation she described was real; it had just not all happened in one place. Carson, who had been arguing for greater control of pesticides, would have been dismayed to find out that in a few short years, her fable had become a disturbing reality. The defoliants sprayed on the jungles and cropland of Vietnam would cause widespread deforestation, water contamination, poisoning of farmland, animal suffering and death, degradation of human health, and fetal deformity. Waugh (2010b) calls the choice to blanket Vietnam with Agent Orange "the first declared war on the environment," and "the world's first planned ecocide, in which entire ecosystems were targeted and destroyed" (p. 118).

While it is now generally accepted that Agent Orange has had an ongoing, devastating effect on human and environmental health, this has not always been the case. Constructing chemical impact, especially visually, is challenging. Toxins often do not look toxic.[1] The clear, innocuous-looking Agent Orange belied its lethality. The physical consequences of exposure to toxins, such as cancer, may not be visible

---

Jennifer Peeples is an Associate Professor at Utah State University.

and the contaminants can exist in the soil, water, or air for years, making it difficult to capture the relationship between toxin and effect. In addition, the uncertainty embedded in arguments over toxins, demonstrated by the difficulty of ascertaining whether *this* particular pollutant is responsible for *this* particular illness, further defies visual representation. And yet we live surrounded by contaminants (Environmental Protection Agency, 2009) in cultures where meaning is most often constructed visually, making it imperative to understand how images function in the representation of toxins (Peeples, 2011).

With that purpose in mind, this essay provides an analysis of the photographs used in mainstream popular media to create the meaning for the dioxin-laden defoliant Agent Orange. In it, I introduce visual narratives,[2] describe Agent Orange's use in Vietnam, and discuss the roles of identity and place in the visual representation of toxins. After comparing the visual narrative to the discursive ones that surround and interact with the images, I explore the national and cultural ramifications of the visual representation of Agent Orange and conclude with an explanation of the construction and function of toxic images.

Over eleven million gallons of Agent Orange were sprayed from 1965 to 1970, covering 10% of South Vietnam's jungles, transportation routes, and cultivated areas in an attempt by the US military to deprive the Vietnamese enemy of jungle cover and food. While Agent Orange is the most recognized herbicide, its use only encompassed about 60% of all the defoliants sprayed in Vietnam by the US military (Shuck, 1987). In addition to its health and environmental impacts, "the herbicide – and its notorious contaminant [dioxin] – is as responsible for changing the perception of chemicals in the U.S. as the explosion in Bhopal, India, in 1984 and the leaking waste dump at Love Canal, N.Y. in 1978" (Hanson, 2008, p. 40).[3] With its continued use as yardstick for establishing toxicity (for example, see Clarke, 2000; Tyson, 1991), its presence in the American imagination (books, products, fonts, and rock bands have all been named "Agent Orange"), and, most importantly, its introduction of "dioxin" to the public lexicon (Hanson, 2008), an analysis of Agent Orange is key to understanding American's perception of toxins and risk. In addition, the importance of examining the photographs of Agent Orange cannot be overstated. According to Kennedy (2008), the images from the Vietnam War era clearly show "an American way of seeing, that is, a way of seeing the world that is visually codified and thematized by the national concerns of the United States." He contends, the "visual legacies of the Vietnam War . . . [are] still being played out within American popular culture" (p. 281).

This project also answers a number of calls for research. In her examination of the contrasting frames used by English and Arabic media coverage of the Afghan War, Fahmy (2010) notes, "little of the work examining the framing of news events has focused on visual images" (p. 697). Looking specifically at environmental communication research, Hansen and Machin (2008) maintain that while much has been written on the ways environmental texts shape public perception, the role of images has been neglected. Furthermore, they argue, "[i]f we wish to understand the discourses presented in the media that might shape public perceptions of the environment and green issues we must also understand how these discourses are realized visually" (Hansen & Machin, 2008, p. 777). Peeples (2011) notes that even less research has examined the visual construction of toxins. Finally, this project extends Seager (1993) and Waugh's (2010b) analyses of war, gender, and the environment.

To gather images of Agent Orange, I searched Lexis/Nexis, SIRS Knowledge Source, EPSCO Host, The Historical Newspapers file for the *New York Times*, and Google Images for any photographs dealing with Agent Orange from 1979 to 2008. I narrowed down the hundreds of articles by focusing on those in which Agent Orange was the primary topic and ones that had a strong visual presence to the piece.[4] I compiled 69 articles containing 184 *toxic images*. I define *toxic images* as *visual representations that are found in print or digital media of people, places, or toxins, which are used to make claims of human-produced contamination causing the degradation of the natural (the body or environment)*. The sources include major newspapers, such as the *New York Times* and the *Washington Post*, popular magazines such as *Life* and *Vanity Fair*, as well as news and scientific magazines like *Newsweek* and *Science*. After organizing the images chronologically, I examined each photograph by noting its content (who and what is pictured, where, and when) and its formal qualities (frame, distance, angle, light, etc.). With primary emphasis on the images, I then reread each article comparing the written and the visual depictions of Agent Orange. As this is a rhetorical analysis, as opposed to a social scientific survey, the intent is not to make definitive statements about the frequency or dispersal of images, but to provide insights into the function of photographs as they construct a public understanding of environmental contamination.

A number of award-winning photographs are included in this analysis. The awards speak both to the proficiency of the photographers and the cultural impact of the images. Pictures of the Year International (POYI; 2012) chose four of the photojournalists to have their photographs reproduced. According to the organizations website, POYI's mission is "to empower the world's best documentary photography, to provide a visual portrayal of society, and to foster an understanding of the issues facing our civilization" (http://archive.poyi.org/about). In addition, in 2008, David Guttenfelder was a finalist for a Pulitzer Prize for his Associated Press photographs of children with Agent Orange'-related abnormalities (http://www.pulitzer.org/bycat/Feature+Photography). And finally, in 2006, James Nachtwey's photographs of Vietnamese victims for *Vanity Fair* earned him a second place prize from the News Photographer magazine in the Enterprise Picture Story category and second place in the Best of Photojournalism contest in 2007 (http://bop.nppa.org/2007/still_photography/winners/OES/79105/152089.html).

In the article accompanying Nachtwey's *Vanity Fair* photographs, Christopher Hitchens (2006) begins by explaining the importance of images for understanding toxicity and health:

> To be writing these words is, for me, to undergo the severest test of my core belief – that sentences can be more powerful than pictures. A writer can hope to do what a photographer cannot: convey how things smelled and sounded as well as how things looked. I seriously doubt my ability to perform this task on this occasion.

> Unless you see the landscape of ecocide, or meet the eyes of its victims, you will quite simply have no idea. I am content, just for once – and especially since it is the work of the brave and tough and undeterrable James Nachtwey – to be occupying the space between pictures. (p. 107)

With deep respect for Hitchens' moving statement, this analysis and others like it contend that the written text and visual images do more than simply stand next to one another. In the case of Agent Orange, I found two distinctive visual threads, the first telling the story of American victims, and later, the Vietnamese. Both showed identity, place, illness, and suffering. The text surrounding the images composed a much different story, allowing for a wider cast of characters, including entities that were not easily represented visually: bureaucracy, science, law, and politics. The written text was also able to articulate the uncertainty that plagues discussions of environmental and human poisoning due to the difficulty of making a causal connection between toxins and health. The words and images each constructed meaning for Agent Orange, told stories that were, at times, complementary and, at others, contradictory, with both influencing their audiences' understanding of the impact of Agent Orange on the places and people that are its victims.

Visual Narrative

Of the ways to represent something, photographs are often touted as truest to their source. An image appears to preserve a moment in time, to document or provide evidence that something existed, or at least that it existed the way it appeared at the instance of photographic capture. Unless the method of production is called into question, images are often seen as objective, as if they can speak for themselves – the axiom of a thousand words. Because words take longer to process and images have a more visceral immediacy (Mendelson & Darling-Wolf, 2009), images are particularly effective at representing loss and emotion (Duffy, 2011), documenting tragedy (Kampf, 2006), visualizing trauma (Duffy, 2011), allowing for vengeance (Bajorek, 2010), picturing war and atrocities (Mendelson & Darling-Wolf, 2009; Parry, 2011), or bearing witness (Noble, 2010).

But, images do not simply represent reality. As Gilligan and Marley (2010) note, "Images are...polysemic. Different audience members may pick up on different elements within the same image, or imbue the same element with a different meaning" (sec. 6.1), though the audience is not alone in the process of visual meaning-making. The construction of the image influences the viewer's understanding of the subject. Beyond the question of what should or should not be included in the photograph, the framing of the subject can be altered based on whether it is shot from above or below, from a close distance or from afar, in light or in shadow, or whether the audience is positioned to look straight on or from an angle. Each adjustment made by the photographer influences the audience's feeling of connectivity, of power, or of intimacy with the subject (Kress & van Leeuwen, 2008). Because slight changes can have great impact in audiences' interpretation (see Sturken & Cartwright's, 2009 discussion of the effect of differing hues in the representations of murder suspect O.J Simpson's face), a photographer, a journalist or an editor's choice of one photograph over another for public consumption is neither objective nor inconsequential. "The meaning of an image – or of multiple images, or a sequence of images – is negotiated between the producer(s) of the image and the viewer(s) of the image, and this negotiation is mediated though the image itself" (Gilligan & Marley, 2010, sec. 6.1).

Most importantly, for this discussion, the photojournalists' choices and the audience's predilections do not take place in a vacuum. Both are strongly influenced by the national context in which they are embedded. Photographers construct their images to fit within cultural- and national-specific schema, stereotypes, or narratives that are easily recognizable to their audience, providing "ready-made interpretations" for their viewers. "As images are shaped and read through the lens of these cultural narratives, the subjects of the photographs are positioned in a 'normalized manner'" (Remillard, 2011, p. 130). For example, Kozol (2004), in her analysis of civilian photographs shot during the Kosovo conflict, critiques the simplistic war narrative of "victims, aggressors and rescuers" (p. 5). Along with the subjects of the images, photographs can also situate their viewers. In the pictures of fleeing refugees, the audience is frequently positioned as being "safe," as the refugees are framed moving away from danger and toward the camera (Parry, 2011).

Gilligan and Marley (2010) warn:

> while we certainly cannot fix the meaning of an image we send out into the world, it is surely crucial to be at least aware of how its 'fixed' compositional elements are likely to be understood once it is out there fighting to be heard against its competitor images and the normative contextual backdrop they provide. (sec. 7.2)

Those images that fit within existing narratives or schemas make comprehension and recollection easier for audiences (Fahmy, 2010, p. 699; Mendelson & Darling-Wolf, 2009). Furthermore, the repetition of similar images tends to embed them in the national consciousness – psychologically, emotionally, and symbolically (Hariman & Lucaites, 2007). The relationship between image and culture becomes especially apparent during times of national conflict when visual archives "replenish themselves by employing familiar motifs in new contexts and new motifs in familiar contexts and intersecting with national myths" (Kampf, 2006, p. 265). For example, the structure and symbols of the iconic photograph depicting the flag raising at Iwo Jima was used to establish meaning for the 9/11 tragedy (Hariman & Lucaites, 2007). Fahmy (2010) concludes, "[A]lthough journalists may follow guidelines for objective reporting, different cultural and political perspectives do filter into the newsmaking process, leading to a dominant framing of the news event to the target audience" (p. 712). This tendency to construct images that reify hegemonic narratives becomes even more influential when dealing with the ambiguity and uncertainty inherent in toxins.

Images and Toxic Risk

With toxins' tendency toward banality (they don't *look* dangerous) and their frequent invisibility, individuals attend to certain contamination risks not because they are the most dire, pressing, or dangerous, but because they are the most compellingly articulated. In addition, the rhetoric surrounding the cases of contamination does not always rely on factual information or scientific evidence. "[R]isk judgments are, to some degree, a by-product of social, cultural, and psychological influences" (McComas, 2006, p. 76). This is not to say that there isn't a material reality – a physical truth – at the core of every environmental toxin controversy. But "the responsibility for the crisis, its magnitude, and its duration" are all contestable, creating an exigency requiring discourse that crafts scientific data and community concerns into a coherent crisis narrative (Millar & Heath, 2004, p. 5). Those fabricating the narrative make visual and discursive choices that construct meaning for the public. In doing so, they establish protagonists and antagonists, the nature of the crisis, its origin and duration, where it takes place (at the factory, in Congress, in the soil, in the body), and what are its effects. Rhetors use these narratives to help audiences understand the social and political worlds in which they live. They also "sanction some kinds of actions and not others" (Denton, 2004, p. 7). As with photographers, rhetors may rely on established narrative templates, such as the tragedy, the comic form, or pastoral betrayal (Buell, 1998), to "make risk scenarios intelligible to the reader or viewer in a particular way" (Heise, 2002, p. 763). As the risk becomes a public issue, these narratives compete for positions of prominence in face-to-face decision-making and mediated venues.

Visual narratives are particularly instrumental in crafting a public understanding of risk. Ferreira (2004), speaking specifically of film, maintains that the "medium encourages audiences to identify with the human, social, and environmental conflicts portrayed in the narratives and thus induces vicarious experience of risk events" (p. 200). In addition:

> [m]uch of people's knowledge about the world is gathered from the reading of visual images in television and newsprint media. This is especially so in the case of 'risk events,' from epidemics to ecological disasters, creating a heightened awareness of the fragility of life-systems in face of different kinds of hazards (Ferreira, Boholm, & Löfstedt, 2001, p. 283)

Simply speaking, risk is constructed most potently through a visual medium, wherein people come to conclusions about what they know based on what they see. Since most people in the USA will never have a direct encounter with Agent Orange, and because of the complexity of the science and legality of the toxin, visual representations become the most influential ways in which people came to understand the environmental, physical, and national ramifications of spraying Agent Orange in Vietnam.

Agent Orange

Of the many herbicides used in Vietnam, the most notorious was Agent Orange, named after the band of color found on its barrels. The two goals of "Operation Hades," later given the more innocuous title "Operation Ranch Hand," were to deprive the North Vietnamese of jungle cover and, later, to destroy the crops of the villages suspected of aiding the enemy (Shuck, 1987). Starting in 1964, the defoliant was sprayed from airplanes on which were written, "Only You Can Prevent a Forest." According to US and Vietnamese military reports, the number of people directly sprayed was a little over 3000, with 2.1 million to 4.8 million indirectly affected (Stellman, Stellman, Christian, Weber, & Tomasallo, 2003).

As early as 1952, army officials had been informed by Monsanto Chemical Company, a producer of Agent Orange, that the defoliant was contaminated with a toxic substance that raised concerns for human health, especially when applied at high concentrations by inexperienced operators (Shuck, 1987). By the late 1960s, the source of danger was found to be a contaminant created in the production process (2,3,7,8 tetrachlorodibenso-p-dioxin), which has been described as "perhaps the most toxic molecule ever synthesized by man" (Shuck, 1987, p. 18).

With growing public concern, the government officially halted the application of Agent Orange in 1971 (Stellman et al., 2003). The initial debate over the use of chemical defoliants in Vietnam became a vociferous fight for reparations for soldiers victimized by those choices. In 1984, US veterans were awarded an out-of-court settlement of 180 million dollars from the chemical companies who produced Agent Orange, the largest compensation package in US history at that time. At no point, however, did the chemical companies or the US federal government admit legal liability. One of the primary reasons no one was forced to claim responsibility is the difficulty of establishing a causal relationship between toxins and human health. With uncertainty playing a central role in both the trial and the surrounding discursive narrative, Shuck's description of the difficulty of creating a definitive link between Agent Orange and the devastating human effects is particularly telling:

> Information on the dioxin levels in the Agent Orange to which American soldiers were exposed was even less reliable. The areas in which aerial spraying missions had been conducted were known, but it was impossible to know how much of the herbicide reached the ground where people could ingest or inhale it; the levels to which specific individuals had been exposed was impossible to learn. [A]lthough dioxin at certain levels was clearly capable of causing serious diseases, those same diseases could also result from other causes. It followed, then, that inferences concerning Agent Orange's causation of particular diseases in particular individuals would remain weak and speculative. (Shuck, 1987, p. 18)

In 1994, the Veterans Benefits Improvement Act formally established the presumption that twelve diseases are linked to Agent Orange: chloracne, diabetes, peripheral neuropathy, porphyria cutanea tarda, Hodgkin's disease, non-Hodgkin's lymphoma, chronic lymphocytic leukemia, soft-tissue sarcoma, respiratory cancers, multiple myeloma, and prostate cancer (Schensul, 2005). The January 2007 issue of *Science* added that there is "limited or suggestive evidence of an association" for spina bifida in offspring of exposed individuals (Stone, inset), with spina bifida being the only birth defect recognized as a result of Agent Orange exposure.

Claiming many of the same ailments as the US veterans and dealing with the same uncertainties, Vietnamese victims of Agent Orange have been unsuccessful in their lawsuits against US chemical companies. That fact that Agent Orange was sprayed during war also changed the rules of legal liability. Though horrific in its effects, US federal judge, Jack B. Weinstein declared that the use of Agent Orange was not an act of chemical warfare. In his decision against a damage suit filed on behalf of millions of Vietnamese, he wrote, "The prohibition of [the use of poison] extended only to gases deployed for their asphyxiation or toxic effects on man, not to herbicides designed to affect plants that may have unintended harmful side effects on people" (Glaberson, 2005, p. B6). Goodman, a lawyer for the Vietnamese claimants countered, "He ruled as a matter of law that what these defendants manufactured was not a poison, whereas even these manufacturers recognized that it was at the time" (Glaberson, 2005, p. B6). During the same period as the legality of Agent Orange was being repeatedly debated in the courtroom, the meaning of Agent Orange was being constructed for the American people through visual and written narratives found in US newspapers and magazines.

In the following sections, I describe the two distinct visual narratives that make up the story of Agent Orange. As explained previously, images are polysemic (Gilligan & Marley, 2010, sec. 6.1). Different audiences perceive and emphasize different aspects of a photograph, creating variant readings of an image, though they are somewhat contained within a shared cultural or national context. My reading of this collection of images emphasizes identity (specifically gender), place, and toxin. I offer that other valid readings of Agent Orange are possible as the viewer, image, focus, and/or context changes. Next, I compare my reading of the visual narrative to the discursive narratives found in the article's text. I then discuss the national and cultural ramifications of these Agent Orange narratives, and conclude with an explanation of the construction and function of toxic images.

Bringing Agent Orange home

The first narrative of Agent Orange is an American story. Though spraying took place in the jungles and rice paddies of Vietnam from 1964 to 1971, there is only one picture, reproduced twice, of Southeast Asia (Severo, 1979a; Shear, 1979). Equally surprising, during the same time period, I found no photographs of Vietnamese people in relation to Agent Orange. While the written narrative locates Vietnam as the site of contamination, with each soldier's story beginning with a discussion of being directly sprayed or walking through the dioxin-laden foliage in Vietnam, the scene for the visual narrative is the USA.

Perhaps less surprising, considering that the vast majority of Americans who were sprayed with Agent Orange were soldiers, all the photographs, besides two in one article, show men as protagonists (Dullea, 1981, p. A1). In the images of Americans, the male protagonists are constructed into three categories: victims, fathers, and soldiers-activists. As with other representations of soldiers at the time, images of "Vietnam soldiers focus[ed] almost exclusively on the more intimate, individual or collective of separate voices" (Jeffords, 1989, p. 2).

The primary way the men are shown in print media is as victims of the defoliant. The men are pictured with neck braces, crutches, wheelchairs, emaciated bodies, and swollen and deformed hands. As a viewer, it is impossible to tell which of the injuries are from the physical war and which are from the chemical. Victims with invisible ailments are shown, too, their illnesses made evident through their gaze and identification with the audience. The victims sit or lean – none are photographed in motion. They stare straight out at the viewer with grave faces (for example, Barron, 1980; Severo, 1979b; Taubes, 1988). The description over one of these photographs explains the complicated nature of the soldiers' illness:

> The Government does not acknowledge that the herbicide known as Agent Orange harmed anything but plant life in Vietnam, so the two doctors in the herbicide clinic cannot do much more than offer sympathy when war veterans insist that their rashes, headaches and more serious troubles were caused by the defoliant. (Barron, 1980, p. A19)

In contrast to the pictures of individuals with visible ailments, who avert their eyes from the viewer and present their bodies for the audiences' appraisal, these photographs are "demand" shots in which the subject makes direct eye contact with the viewers, implicitly asking something of them (Kress & van Leeuwen, 2008). Because of the uncertainty surrounding toxin-related illnesses, especially at that time, the men may have been framed as asking the viewer to believe their ailments and suffering were real, even when science and the government did not recognize their existence or cause. The interaction of image and text raises some of the same tensions of what is known and what is unknown that can be found in images of toxic landscapes (Peeples, 2011).

The second way men are constructed in the American visual narrative is as fathers. Of the eight award-winning pictures in Wendy Watriss's (1981) photographic essay "Tracking Agent Orange" in *Life* magazine, three are of fathers holding or touching their children (http://archive.poyi.org/items/show/10690). This intimate and loving interaction between fathers and their families is one that is repeated through this period. In a front page article in the *New York Times*, an image shows a father in the backyard with his two children (Severo, 1979b). The scene is a familiar snapshot. Identification with the traditional family photograph is made strange by the recognition that the father is in pajamas and a robe in his backyard, and he is unable to help balance his son on his shoulders or place his arm around his daughter due to the crutches that support him. The fathers in these images are all photographed in intimate, domestic settings: on the floor with the kids in the family room, sitting in the kitchen, or standing in front of their homes.

Women play a minor role in this visual narrative. When mothers are included, such as in Watriss's photograph of the family around the kitchen table, they are off to the side of the frame not touching or interacting with the children. This contravention of a traditional family role is even more striking when the viewer recognizes that the children too are victims of Agent Orange. The daughter being nuzzled by her father does not have an arm, the boy on his father's lap has a large facial tumor, and a child being hugged by his father, as we see in subsequent images, has deformed forearms and hands. The prototype of the nurturing, protective mother, especially to a sick child, is violated by these images, with fathers in their place. The text supporting these photographs explains how each man was contaminated with Agent Orange, and that it is suspected of causing illnesses and deformities in their children. According to Jeffords (1989), the victim images of soldiers are also consistent with other representation of soldiers, following the war. "Rejected by an American society that came to see these men as emblems of loss, moral failure or national decline, Vietnam representation could effectively portray them as 'victims' of society, government and the war itself'" (Jeffords, 1989, p. xiv); and in this case, victims of contamination caused by the same powerful triad.

In contrast to the father and victim images, the pictures of soldiers and activists document men taking action in public places. The images are of protests, men talking to reporters, testifying in court, and politically organizing (Blumenthal, 1984; Latimer, 1981). The uniformed soldier with a cane is in midstride, walking down the street (Blumenthal, 1984). Protesters are seen demanding rights for Agent Orange victims in marches and sit-ins. None of the father/victim images show soldiers in uniform. None of the soldier/activist photographs are posed or have the subject looking at the camera. In a more traditional masculine frame (Kilbourne, 2010), the soldier/activist photographs appear to be offering the viewer their actions (not themselves) for audience appraisal.

At the conclusion of the veterans' trial in which the chemical companies agreed to pay 180 million dollars to the twenty thousand vets who were experiencing illness, disease, and death from Agent Orange, the lawyer Victor John Yannacone Jr. announced the end of the Vietnam conflict was not the pulling out of Saigon in 1975, but the conclusion to the trial almost a decade later, stating, "The veterans have won the final battle of the Vietnam War" (Blumenthal, 1984, p. A1). But that would not prove to be the case. The battle over the meaning of the war would continue for years to come, including the continued fight for reparations for those victimized by the decision to spray Agent Orange. It would also take another decade before the Agent Orange visual narrative would be complete, bringing in a victim more imperiled than their American counterparts, providing the opportunity for a reframing of the role of the American soldier in the national, visual narrative and move the site of contamination back to Vietnam.

The Vietnamization of Toxic Risk

While pictures of American victims of Agent Orange were making front pages and were the subject of photographic essays in the USA, Jones Griffiths (2003) was unable to sell any of his pictures of the afflicted Vietnamese for almost two decades (The Digital Journalist http://www.digitaljournalist.org/issue0401/griffiths_intro.html). The photojournalist, renowned for his book of war photography *Vietnam Inc.*, would later compile his images into an award-winning tome: *Agent Orange: "Collateral Damage" in Viet Nam*. The disturbing nature of the photographs, the raw wound left by the end of the war, and the lingering prejudice against the Vietnamese "enemy" may have been instrumental in the delay. As Fahmy (2010) explains, "The framing of suffering occurs in a dichotomy that makes distinctions between the unworthy victims depicted as enemies and the worthy victims who suffer" (2010, p. 700). But once the photographs of the Vietnamese visual narrative were in circulation in newspapers and magazines, the startling images of children with grotesque deformities, the renewed questioning of the connection to Agent Orange, and in 2004, a legal battle, this time waged by Vietnamese victims, created a gestalt shift in the larger Agent Orange narrative.

Roughly half of the images I collected for this time period were photographs of Vietnamese children with obvious physical and mental disabilities. Missing limbs, hydrocephaly, cleft palates, fused eyelids, deformed and twisted bodies, and vacant stares fill the frames of these images. The children are most often pictured alone, in a crib, on the floor, in a bed, with no parent, or caregiver in the frame. Without question, the images are visually disturbing spectacles. While many of the illnesses of the first generation of victims do not show (such as diabetes or cancer), the effects of Agent Orange on the second and third generation is shockingly visible. Equally important in terms of narrative potency, Buell (1998) argues that a heightened level of persuasion occurs when the victims appear to have no choice. Children are perceived to be the most innocent and powerless of victims.

In direct contrast to the American narrative, when Vietnamese children are photographed with another person, it is almost always with the mother or female caregiver. The women are shown holding, bathing, and feeding the children at home (http://www.vanityfair.com/politics/features/2006/08/hitchens200608). In the photographs, the mothers are consistently portrayed as passive, appearing helpless to the circumstances that have befallen them. With downcast or horizon-gazing eyes, the central feature of the mothers' faces is despair.

These photographs may resonate with the viewer because of their similarity well known to Western iconographic images. From Dorthea Lange's Migrant Mother to the Pieta, these photographs are recognizable in their subject matter and composition. And "while images must in some sense be familiar, they must also contain an element of distinctiveness and be dramatized in a symbolic way if they are to make a major impact" (Ferreira, Boholm, & Löfstedt, 2001, p. 284). Western audiences can instantly identify the commonplace image of a struggling mother and an endangered child, but it is the deformity of the child and the sights of non-Western poverty (woven floor mats for sleeping, conical hats, bare walls, dirt yards) that make the Vietnamese photographs' narrative distinct and compelling.

It is in the Vietnamese visual narrative that the identity of the American soldier is reframed. While the discursive narrative continues to discuss the health issues of the American service men and their families, the soldiers and their illnesses all but disappear from the Vietnamese visual narrative. In the 108 print pictures in this narrative, only two are of Americans still dealing with the effects of Agent Orange. The choice to exclude the continued health problems of American veterans and their children is made apparent in Nachtwey's (2006) critically acclaimed photo essay in *Vanity Fair* (http://www.vanityfair.com/politics/features/2006/08/nachtwey_photoessay200608). The magazine version of this essay has eleven images, all of the second generation of afflicted Vietnamese. The online version has 21 images: 14 of

Vietnamese and 6 photographs of American victims, including the now-bedridden soldiers and their disabled adult children. One image shows a six-year-old boy stricken with spinal bifida whose grandfather had been sprayed with Agent Orange in Vietnam. Not one of the photographs of Americans was included in Nachtwey's print essay. In the transition from the American narrative to the Vietnamese, the bodies at risk changed from the American soldiers to Vietnamese children. The site of contamination too moves from US domestic spaces to the unfamiliar and distant sites of Vietnam.

Science, Uncertainty, and Blame

The written text surrounding the toxic images tells a different story from its visual counterpart. In the American narrative, four protagonists battle to create the meaning of Agent Orange for the reading public: the chemical industry (especially the seven chemical companies that ended up settling in the class-action lawsuit), the scientists tasked with finding the "truth" about Agent Orange exposure and human health, the federal government (primarily the Veterans Administration and the Center for Disease Control), and the American veterans characterized by their illness and their anger at the lack of support and compensation. Another player in the story was Agent Orange itself, often characterized in terms of its chemical composition, level of dioxin, as well as its means and volume of dispersal.

The written text centers on the lawsuit raised by American soldiers suing chemical companies for the health effects they experienced from their exposure to Agent Orange. Following the lawsuit, the story switches to one of fair distribution of compensation. Because the standard for holding the chemical company responsible for veterans' illnesses was causal, the discursive narrative is replete with discussions of the difficulty of "proving" Agent Orange's toxicity for human health. Competing science and scientists, the inability to state without a doubt the health effects of environmental toxins, and the federal government's actions/inactions in funding scientific studies and supporting ailing soldiers dominated the narrative. An underlying current of blame (the Veterans Administration, the Center for Disease Control, the federal government, the "system," the chemical companies, etc.), liability, and conspiracy (burying evidence and ignoring scientific findings) also motivate the narrative. The visual narrative does not allow for blame or the details of conspiracy, focusing the attention on effect.

The written text surrounding the Vietnamese images has three primary characters: the first-generation victims with Agent Orange diseases, their children and grandchildren (both alive with physical and psychological deformities, and those who were miscarried or died at birth), and science. Unlike the Vietnamese visual narrative, the discursive narrative gives relatively equal time to the Vietnamese veterans' illnesses and the damage to the second and third generations. Each story of an individual follows a similar script: the person explains where and when he/she came into contact with the defoliant, how Agent Orange smelled or tasted, what the spray did to the surrounding environment (blackened plants, dead birds, oily looking water), actions taken that may have exacerbated their exposure (drank or swam in the water, ate the fish), and the physical and psychological ramification of that contact to themselves and their families (illness, death, cancer, multiple miscarriages, stillbirths, and defects; for example, see Chandrasekaran, 2000).

Science is clearly the protagonist in both the American and Vietnamese discursive narratives, in that it is portrayed as the entity that would provide knowledge and insight into the chemistry of the defoliant and the truth of its human/environmental effects. The visual narratives do not include science in that it does not have a clear material presence that can be easily articulated in an image, though the inclusion of a graph or chart explaining findings can be seen as a symbolic marker for the visually amorphous "science."

The Vietnamese visual narrative constructs Agent Orange with a much simpler plot. In addition to marginalizing science, there is no uncertainty. The slipperiness of language that allows for allegations of Agent Orange damage to be attributed to a source from "the Vietnamese government" or to be tamed with expert assertions of a lack of proof does not exist in the visual narrative. For a comparative example, *Science* magazine in 2007 examined "Agent Orange's Bitter Harvest." The third paragraph carefully states:

> Vietnam claims that the children's disabilities were caused by parental exposures to Agent Orange. Western scientists have long been at odds with their Vietnamese counterparts over the strength of evidence correlating exposure to dioxin – a toxic contaminant of the herbicide – and illnesses in individuals, particularly birth defects. "The Vietnamese government is using malformed babies as a symbol of Agent Orange damage," says Arnold Schecter, a toxicologist at the University of Texas School of Public Health in Dallas, who remains cautious about making associations after studying Agent Orange for more than 20 years. (Stone, 2007, p. 176)

The Vietnamese visual narrative offers no such qualifiers. In order, the first three pictures in the article show (1) Agent Orange being sprayed over the rice paddies, (2) the denuded landscape, and (3) young people with deformities. While most of the articles from the sample did not contain all the three pictures in this particular order, the visual narrative read across multiple sources consistently shows images of spraying and images of deformity, making a correlation, if not a causal argument, between the two in the mind of the viewers. The visual narrative is, therefore, capable of providing clarity to the complicated and contested nature of toxic issues that would be considered controversial if stated in words.

National, Cultural, and Environmental Significance

Photojournalism is plagued with questions of its impact for those who suffer in front of the camera. Noble (2010) contends that "icons of outrage...starving child, the destitute earth quake survivor, the victim of torture...'may stir controversy, accolades, and emotion, but achieve absolutely nothing'" (pp. 187–188). It took 30 years for the Vietnamese victims to get their day in court and even then they did not prevail. A person could easily argue that the visual and discursive narratives of Agent Orange failed to alleviate the misery for its greatest victims and, therefore, lacked impact. That said, Noble (2010) argues that a limited outcome for the image's subjects should not preclude us from analyzing how photographs work in other capacities. In this case, the overarching Agent Orange narrative, the one that contains both the American and Vietnamese stories, has altered its potential for inclusion into the national movement against toxins, and its gendered storyline may have had an impact on the American national and cultural understanding of toxic risk.

Even before the loss of the Vietnam War, American masculinity was being tested. The demand for rights for women, people of color, and college students taking place in the 1950s, 1960s and 1970s shifted the "stability of the ground on which patriarchal power [had rested]" (Jeffords, 1989, p. xi–xii). The Vietnam War further challenged masculinity for the US soldiers. The American visual narrative shows evidence of this crisis of masculinity. Instead of being caused by the loss of the war or the failure of the US government, it was the toxic agents that were emasculating men who were once strong and vital. The arrival in the Vietnamese narrative of a new and even more powerless casualty of Agent Orange, the deformed Vietnamese child, would significantly alter the American's role in the toxic narrative, removing the Americans/soldiers from the primary role of "victim." At the same time, the role of the American soldier was being transformed, the ongoing pain of the American victims and their children disappears from the visual narrative almost entirely. In addition, the place of contamination is no longer the American home, though the toxins/effects still reside in the bodies and genes of the soldiers and their families, but is visually reassigned back to Vietnam.

The shifting of risk from domestic to distant places can also be seen in a discrepancy between the visual and written narratives. Dioxin poisonings in the USA were sporadically included in discursive descriptions of Agent Orange: spraying cattle, clearing rail lines, and roadways (specifically around Alsea, Oregon), illnesses experienced by Kansas farmers, and the dioxin contamination in Times Beach, Missouri, and in and around Monsanto and Diamond Alkali plants (all three directly or indirectly linked to Agent Orange production). In the 184 pictures I covered in this analysis, only two images of domestic contamination were included (Mansnerus, 1998; Severo, 1979a), making them outliers in the visual narrative of Agent Orange and dioxin.

Looking at the shift in the Agent Orange narrative, one could rightly argue that the discursive and visual transference back to Vietnam was warranted. A greater number of people there were harmed by the herbicide, Agent Orange, which is still present in the soil of Vietnam (especially around old US air bases), and effects have been even more horrific due to the amount of defoliant and length of exposure. That acknowledged, the visual exclusion of the ill American veterans, their children and grandchildren, and the contaminated places in the USA, may have allowed for a belief that while there may be a risk of toxins "here," it is insignificant compared to what we are seeing over "there." This construction of risk that allows Americans' to feel a sense of toxic immunity (or at least resistance) is contingent on the presence of a distant, weak, feminized other, which may have influenced how Americans' conceptualized and made decisions concerning toxins. No *environmental* policy altering the production, use, or distribution of toxins or defoliants came in response to Agent Orange.[5] This is perhaps more surprising when compared to the banning of DDT that followed the publication of the book *Silent Spring*.

The visual disappearance of the American, domestic victims of Agent Orange may have also lessened critics' abilities to tie Agent Orange and its victims into the national environmental consciousness. Waugh (2010b) argues that the racist and imperialistic rationalizations for the use of Agent Orange make it one of the "greatest environmental injustices of the twentieth century" (p. 115), yet it is all but missing from the US environmental history. In a review of 17 environmental/environmental movement texts including Merchant (2007), Warren (2003), Sale (1993), and Brulle (2000), I found one reference to Agent Orange, in which Barry Commoner is quoted as having said that relaxing environmental regulations would weaken the claims of American Veterans exposed to Agent Orange (Egan, 2009).[6] Though happening in overlapping time periods and dealing with the similar issues of toxicity, gender, and uncertainty (Peeples & DeLuca, 2006), the environmental justice and antitoxins movements do not include the ongoing plight of the contaminated US soldiers and the Vietnamese people, prohibiting the expansion of resources, knowledge, and global reach that that the union might have allowed.

Toxic Images

The previous sections have argued for how Agent Orange images function with a specific emphasis on narrative, culture, place, and identity. In this last section, I draw the broader conclusions from these discussions and apply them to an examination of toxic images. With a seemingly unending stream of new chemicals being used in previously unthinkable ways, it is important to understand how images visually construct contamination both for critical analysis and political activism.

Toxic Images are Intrinsically Tied to their Larger Cultural and National Contexts

Images are the primary means of representing national or cultural hegemonic perspectives. Fahmy argues, "Because of their ability to succinctly encapsulate culture, subject, audience, while hiding the decision-making involved in production of the image, photographs are especially adept at representing and reifying national ideologies" (2010, p. 698). This is true for all photographs, but, I would argue, it is especially important to note for toxic images. Many of the most potent and destructive toxins are invisible or dull, making them seem unremarkable. For toxins to have impact or be memorable, they need to be situated within a larger narrative, as they often do not have the visual presence to create significance on their own. In analyzing the visual construction of contamination, one must be cognizant of the larger cultural story that is being told by and through the toxic images, stories of national identity, of culture, of gender, of race, and, most significantly, of power.

The Meaning of a Toxic Image is Constructed through its Visual Narrative

As opposed to time- and space-specific catastrophes (tornados, hurricanes, fires, or famines), toxins are not so easily contained. They move between elements (water, air, soil, bodies) and are transported to sites far from their production. Some chemicals change their structure, depending where they are located in the food chain, and as we saw with Agent Orange, physical deformities can look different in the bodies of first and second-generation victims. The uncertainty surrounding the effects of toxins makes them even more difficult to visually capture. To create a compelling argument for contamination requires a toxin, a victim, and a site (Ferreira, Boholm, & Löfstedt, 2001), which, as we saw, may not be on the same continent, much less contained within the same photographic frame. Multiple images are often necessary to create the "story" of contamination.

Most importantly, an understanding of toxins comes through an examination of the relationship between photographs, in that the effects of contaminants are found causally. Too much fertilizer causes algae blooms, mercury emissions cause lower IQs, and spraying Agent Orange causes reproductive deformities. Single images are ill equipped to make the causal arguments necessary to establish contamination.

Images Allow "Testimony" from Victims Who May Not Be Able to Express Their Suffering through Language

In an interview, the photographer Grossfeld (2002) described how he chose what to photograph when covering environmental issues:

> I listened to what scientists observed was happening, but I kept my camera's eye fixed on the haunting faces of children... Their expressions and circumstances bespoke the consequences of the environmental tragedies in ways that any retelling of the experts' verbal arguments never could. (p. 42)

Especially important for environmental poisoning, images can give an opportunity for people to be involved in the social construction of a toxin, who might otherwise be excluded due to language, translation, marginalization, and/or reduced capacity due to dioxin's manifestation in physiological and psychological abnormalities. Images may then allow victims to participate in ways that written discourse does not, though as with all images, the agency of the subject is always partial as the meaning of the image is negotiated between photographer, subject, and viewer.

Images Obfuscate the Scientific, Political, and Environmental Complexity of Toxins

Images and the discursive text that surrounds them often do not tell the same story. In their analysis of readers' interpretations of images and text, Mendelson and Darling-Wolf (2009) found that photographs can overwhelm and simplify a "more complicated reality posed by the text." (p. 801). They argue:

> On a basic level, words are a symbolic sign system, while photographs are more iconic and indexical... Words take more time to process, perhaps leading to a

more reasoned response, while viewers' response to pictures tends to be more immediate and emotional. Photographs are more concrete, less abstract, than words, tied more closely to a specific time and place. This is not to suggest that photographs cannot represent more abstract concepts, but that they are more grounded in the denotative. (p. 801)

With scientifically, politically, and culturally complex subjects like environmental contamination, the ease, immediacy, and emotional resonance of the images can overpower the written text. Amorphous concepts like science or complex bureaucracies like the Veterans Administration may be missing from the visual narrative or represented synecdochically through a diagram or an "expert" shot. A written narrative allows for a broad definition of "protagonist" or "antagonist" (dioxin, science, government, etc.), with those roles most consistently being filled in the visual narrative with individuals or places that stand in to represent less visual entities.

Images Can Lessen the Uncertainty Inherent in Discussions of Contamination

The simplification of toxicity that takes place in images may also function positively to clarify risks that have been purposefully obfuscated, jargon-laden, or complex beyond the grasp of an interested public. As was seen in the article in *Science*, the uncertainty of the connection between Agent Orange and birth defects found in the written explanation disappeared in the visual narrative.

Finally, as this analysis extrapolates from one case study to a larger discussion of toxic images, it is imperative for the next stage of research to examine whether the observations found here are consistent with other examples of contamination and risk. Unfortunately, for all of us, there seems to be no end to the number and variety of these examples for further study.

Notes

[1] For an excellent discussion of the limits of the visual for representing toxins, see Pezzullo (2007).

[2] I define a visual narrative as a subject-specific sequence of images found within and across media during a defined time period.

[3] Though the contamination of the homes at Love Canal is more frequently mentioned in terms of impact on the nation's environmental awareness, Agent Orange has received a greater amount of media attention and for a longer period of time. Searching ProQuest Newsstand, from 1980 to 2012, there were 3094 articles with "Agent Orange" in the title as compared to the 916 articles 1978–2012 with "Love Canal" in the title. Perhaps even more telling, since 2010 there have been 361 Agent Orange articles and only 10 focused on Love Canal (again searching article titles for the key terms). Unfortunately, for this analysis, articles from ProQuest are primarily saved as "full text" documents, which exclude the pictures from the original article. In terms of cultural impact (a difficult concept to quantify) in the 178 articles from the *New York Times* from 1979 to 2008 (again with Agent Orange in the title), 19 articles (10.7%) were the lead for their section (A1, B1, C1 or on the first page of a special edition).

[4] Archival photographs are difficult to obtain. Articles posted through online databases may not include the images found in the original text, some are under separate copyright and are

inaccessible, and many of those found on microfiche or microfilm are of poor quality, excluding them from a visual analysis.

[5] That is not to say that there were no new policies that came in response to Agent Orange, but the impact was focused on changes in military protocol and soldier health. President Ford signed an executive order prohibiting the military use of defoliants, the Senate ratified the Geneva protocol banning chemical weapons, and President H. W. Bush signed Agent Orange Act of 1991, authorizing a long-term health study on soldiers who were exposed to the defoliant (Waugh, 2010a).

[6] For a complete list of texts, please email the author.

References

Bajorek, J. (2010). Introduction: Special section on recent photography theory: The state in visual matters. *Theory, Culture and Society, 27*, 155–160. doi:10.1177/0263276410383719

Barron, J. (1980, March 18). Veterans criticize Agent Orange tests. *New York Times*, p. A19.

Blumenthal, R. (1984, May 8). Veterans accept $180 million pact on Agent Orange. *New York Times*, p. A1.

Brulle, R. (2000). *Agency, democracy, and nature: The US environmental movement from a critical theory perspective.* Cambridge, MA: MIT Press.

Buell, L. (1998). Toxic discourse. *Critical Inquiry, 24*, 639–665. doi:10.1086/448889

Carson, R. (1962). *Silent spring.* Boston: Houghton Mifflin Company.

Chandrasekaran, R. (2000, April 18). War's toxic legacy lingers in Vietnam: Cancers, birth defects attributed to US use of Agent Orange. *Washington Post*, p. A1.

Clarke, K. (2000, June). Future under fire: Depleted uranium munitions could be the next Agent Orange crisis. *US Catholic*, p. 23.

Denton, R. E. Jr. (2004). *Languages, symbols, and the media: Communication in the aftermath of the World Trade Center attack.* New Brunswick: Transaction Publishers.

Duffy, J. (2011). In the wake of trauma: Visualizing the unspeakable; unthinkable in Marie Darrieussecq and Hélène Lenoir. *Word & Image, 27*(4), 416–428. doi:10.1080/02666286.2011.627215

Dullea, G. (1981, March 23). Women speaking out on the effects of duty in Vietnam: Like men who fought, they tell of anxiety and painful recall. *New York Times*, pp. A1, B12.

Egan, M. (2009). *Barry Commoner and the science of survival: The remaking of American environmentalism.* Cambridge MA: MIT Press.

Environmental Protection Agency. (2009). *Toxics Release Inventory (TRI): National analysis overview.* Retrieved from http://www.epa.gov/tri/tridata/tri09/nationalanalysis/overview/2009TRINAOverviewfinal.pdf

Fahmy, S. (2010). Contrasting visual frames of our times: A framing analysis of English- and Arabic-language press coverage of war and terrorism. *International Communication Gazette, 72*(8), 695–717. doi:10.1177/1748048510380801

Ferreira, C. (2004). Risk, transparency and cover up: Media narratives and cultural resonance. *Journal of Risk Research, 7*(2), 199–211. doi:10.1080/1366987042000171294

Ferreira, C., Boholm, A., & Löfstedt, R. (2001). From vision to catastrophe: A risk event in search of images. In J. Flynn, P. Slovic, & H. Kunreuther (Eds.), *Risk, media and stigma: Understanding public challenges to modern science and technology* (pp. 283–300). London: Earthscan Publications Ltd.

Gilligan, C., & Marley, C. (2010). Migration and division: Thoughts on (anti) narrativity in visual representation of mobile people. *Forum Qualitative Social Research 11*(2). Retrieved from http://www.qualitative-research.net/

Glaberson, W. (2005, March 11). Civil lawsuit on defoliant in Vietnam is dismissed. *New York Times*, p. B6.

Griffiths, P. J. (2003). *Agent Orange: "Collateral damage" in Vietnam*. London: Trolley Ltd.

Grossfeld, S. (2002, Winter). Connecting the human condition to environmental destruction. *Nieman Reports*, pp. 42–44.

Hansen, A., & Machin, D. (2008). Visually branding the environment: Climate change as a marketing opportunity. *Discourse Studies, 10*, 777–794. doi:10.1177/1461445608098200

Hanson, D. (2008, March 17). Agent Orange's legacy: The battle over dioxin and reputed health problems shaped public perception of chemicals. *Chemical & Engineering News*, p. 40.

Hariman, R., & Lucaites, J. L. (2007). *No caption needed: Iconic photographs, public culture, and liberal democracy*. Chicago: University of Chicago Press.

Heise, U. (2002). Toxins, drugs, and global systems: Risk and narrative in the contemporary novel. *American Literature, 74*, 747–778. doi:10.1215/00029831-74-4-747

Hitchens, C. (2006, July 24). The Vietnam syndrome. *Vanity Fair*, pp. 106–111.

Jeffords, S. (1989). *The remasculinization of America*. Bloomington, IN: Indiana University Press.

Kampf, Z. (2006). Blood on their hands: The story of a photograph in the Israeli national discourse. *Semiotica, 162*, 263–285. Retrieved from http://www.xolopo.com/linguistics_and_literature/blood_hands_story_a_photograph_israeli_national_9991.html

Kennedy, L. (2008). Securing vision: Photography and US foreign policy. *Media, Culture & Society, 30*(3), 279–294. doi:10.1177/0163443708088788

Kilbourne, J. (2010). *Killing us softly 4* [Motion picture]. Northampton, MA: Media Education Foundation.

Kozol, W. (2004). Domesticating NATO's war in Kosovo/a: (In)Visible bodies and the dilemma of photojournalism. Meridians: Feminism, Race. *Transnationalism, 4*(2), 1–38. Retrieved from http://search.ebscohost.com.dist.lib.usu.edu/login.aspx?direct=true&db=aph&AN=13006588&site=ehost-live

Kress, G., & van Leeuwen, T. (2008). *Reading images: The grammar of visual design* (2nd ed.). London: Routledge.

Latimer, L. (1981, June 25). 8 Vietnam veterans fast here to get action on Agent Orange. *Washington Post*, pp. D1, D4.

Mansnerus, L. (1998). Newark's toxic tomb. *New York Times*, Sec. 14, pp. 1, 13.

McComas, K. (2006). Defining moments in risk communication research: 1996–2005. *Journal of Health Communication, 11*, 75–91. doi:10.1080/10810730500461091

Mendelson, A., & Darling-Wolf, F. (2009). Readers' interpretations of visual and verbal narratives of a *National Geographic* story on Saudi Arabia. *Journalism, 10*(6), 798–818. doi:10.1177/1464884909344481

Merchant, C. (2007). *American environmental history: An introduction*. New York: Columbia University Press.

Millar, D. P., & Heath, R. L. (2004). *Responding to crises: A rhetorical approach to crisis communication*. Mahway, New Jersey: Lawrence Erlbaum Associates.

Nachtwey, J. (2006, July 24). The Agent Orange syndrome. *Vanity Fair*, pp. 106–111. Retrieved from http://www.vanityfair.com/politics/features/2006/08/nachtwey_photoessay200608#intro

Noble, A. (2010). Recognizing historical injustice through photography: Mexico 1968. *Theory, Culture & Society, 27*(7–8), 184–213. doi:10.1177/0263276410383714

Parry, K. (2011). Images of liberation? Visual framing, humanitarianism and British press photography during the 2003 Iraq invasion. *Media, Culture & Society, 33*(8), 1185–1201. doi:10.1177/0163443711418274

Peeples, J. (2011). Toxic sublime: Imaging contaminated landscapes. *Environmental Communication: A Journal of Nature and Culture, 5*(4), 373–392. doi:10.1080/17524032.2011.616516

Peeples, J., & DeLuca, K. M. (2006). The truth of the matter: Motherhood, community and environmental justice. *Women's Studies in Communication, 29*, 39–58. Retrieved from http://search.ebscohost.com.dist.lib.usu.edu/login.aspx?direct=true&db=aph&AN=20954727&site=ehost-live

Pezzullo, P. (2007). *Toxic tourism: Rhetorics of pollution, travel, and environmental justice*. Tuscaloosa, AL: University of Alabama Press.
Pictures of the Year International POYI. (2012). About. Retrieved from http://poyi.org/67/history.php
Remillard, C. (2011). Picturing environmental risk: The Canadian oil sands and National Geographic. *International Communication Gazette*, *73*, 127–143. doi:10.1177/1748048510386745
Sale, K. (1993). *The green revolution: The American environmental movement, 1962–1992*. New York: Hill and Wang.
Schensul, J. (2005, April 24). A long-ago war's grimmest legacy lives on: Vietnamese still suffer from Agent Orange exposure. *The Record (Bergen County, NJ)*, p. A01. On-Line LexisNexis Academic.
Seager, J. (1993). Creating a culture of destruction: Gender, militarism, and the environment. In R. Hofrichter (Ed.), *Toxic struggles: The theory and practice of environmental justice* (pp. 58–66). Philadelphia, PA: New Society Publishers.
Severo, R. (1979a, May 27). Two crippled lives mirror disputes on herbicides. *New York Times*, pp. A1, A42.
Severo, R. (1979b, May 28). US despite claims of veterans, says none are herbicide victims. *New York Times*, pp. A1, D8.
Shear, J. (1979, November 4). State weighs Agent Orange bill. *New York Times*, pp. 1, 26, New Jersey Weekly edition.
Shuck, P. (1987). *Agent Orange on trial: Mass toxic disasters in the courts*. Cambridge, MA: The Belknap Press.
Stellman, J., Stellman, S., Christian, R., Weber, T., & Tomasallo, C. (2003). The extent and patterns of usage of Agent Orange and other herbicides in Vietnam. *Nature*, *224*, 681–687. doi:10.1038/nature01537
Stone, R. (2007). Agent Orange's bitter harvest. *Science*, *315*, 176–179. doi:10.1126/science.315.5809.176
Sturken, M., & Cartwright, L. (2009). *Practices of looking: An introduction to visual culture* (2nd ed.). New York: Oxford University Press.
Taubes, G. (1988, April). Unmasking Agent Orange. *Discover*, pp. 42–48.
Tyson, R. (1991, March 15). Oil fires threaten health: Kuwait blazes compared to Agent Orange, *USA Today*, pp. A1–A2.
Warren, L. (Ed.). (2003). *American environmental history*. Maiden, MA: Wiley-Blackwell.
Watriss, W. (1981, December 1). Tracking Agent Orange. *Life*, pp. *65–70*.
Waugh, C. (2010a). *Family of fallen leaves: Stories of Agent Orange by Vietnamese writers*. Athens, GA: University of Georgia Press.
Waugh, C. (2010b). "Only you can prevent a forest": Agent Orange, ecocide and environmental justice. *Interdisciplinary Studies in Literature and Environment*, *17*(1), 113–132. doi:10.1093/isle/isp156

Selling Nature in a Resource-Based Economy: Romantic/Extractive Gazes and Alberta's Bituminous Sands

Geo Takach

Recent literature has explored the interplay between the Romantic gaze and the extractive gaze to conclude that in separating people from nature, both gazes function similarly to subordinate the land to human purposes. Such representations may be seen as part of a wider trend in which media visualizations of nature are based on an implicit ideology, tending to perpetuate and justify existing power relations; those visualizations use images which are increasingly abstract or iconic, and which by force of repetition, replace alternative representations and obscure connections to societal processes such as globalism and consumerism. This article takes up and extends that argument to a critical visual discourse analysis of an official place-branding slideshow produced by the Province of Alberta (Canada), which boasts an economy based significantly on producing non-renewable fossil fuels. In examining that slideshow in terms of Romantic/extractive gazes, this study situates Alberta's rebranding on Corbett's continuum of anthropocentric-ecocentric values; interrogates connections among Alberta's rebranding and invisible flows of power at work in broader, underlying societal processes like globalization, Neoliberalism and consumerism; and tests the commonality of the relationship between extractive and Romantic gazes in light of those processes.

Contemporary society expresses itself significantly through images (Mirzoeff, 2009; Gierstberg & Osterbaan, 2002), boosted by instant and international visual communication on the Internet. Thus, the rising conflict between the continuous economic growth mandated by an increasingly global and pervasive market economy and its accelerating toll on nature is also playing out through visual media (e.g., Cammaer, 2009). Images of nature, both pristine and despoiled, have been deployed in service of both pro-development and environmental concerns. Recent work, drawing on scholars such as Cronon (1995) and Morton (2007), has explored the interplay between the Romantic gaze (sanctifying nature as sublime) and the extractive gaze (viewing nature as a resource to be exploited) in the work of Canadian landscape artists, to conclude that in separating people from nature, both gazes function similarly to subordinate the land to human purposes (Hodgins & Thompson, 2011). Such representations may be positioned in a wider trend in which media visualizations of nature are based on an implicit ideology, tending to perpetuate and justify existing power relations (e.g., Berger, 1972; Sturken & Cartwright, 2008). Those visualizations use images which are increasingly abstract or iconic, and which by repetition, "replace other possible representations, particularly those that locate and connect such issues in actual concrete processes such as globalism and consumerism" (Hansen & Machin, 2008, p. 775).

This article takes up and extends that argument to a critical visual discourse analysis of an official place-branding slideshow produced by a Western Canadian province, Alberta, which boasts an economy that is based significantly and lucratively on producing non-renewable fossil fuels. In examining that slideshow in terms of Romantic/extractive gazes, this study will situate Alberta's rebranding on Corbett's (2006) continuum of anthropocentric-ecocentric values; interrogate connections among Alberta's rebranding and invisible flows of power at work in broader, underlying societal processes; and comment on the commonality of the relationship between extractive and Romantic gazes in light of those processes.

Case Study: Alberta Rebranded

An emerging, international flashpoint in the rising tension between economic development on the current, global model and its ecological costs – and visual manifestations of that conflict – occurs in Canada's fourth-largest province in both land area and population. Alberta hosts the world's largest industrial project (Leahy, 2006), the tar sands – also called oil sands but more precisely, bituminous sands – containing the world's third-largest recoverable source of oil (Alberta, 2012a). Unlike conventional oil, this "synthetic" crude cannot simply be pumped out of the ground. Rather, it is surface-mined by stripping out trees and topsoil; digging up the bitumen (in a natural mix of sand, clay and rock) with a vehicle that scoops a hundred tonnes at a time; loading it into the world's largest trucks; feeding it into a crusher which breaks up lumps and removes rocks; and ultimately using hot water to separate out the bitumen. That separation process was pioneered by the publicly-funded Alberta Research Council in the 1920s and rendered possible by several factors: investment from the federal and provincial governments and from private Canadian and American energy interests; technological improvements; further economic efficiencies; the world's ever-increasing demand for energy; and rising oil prices (Chastko, 2004). More recently, bitumen has also been extracted "in situ" through the injection of steam, a newer process for crude located too deep for surface mining.

Although bitumen extraction has been underway on a commercial scale since 1967 (Clarke, 2008), its global profile has risen since the US included the resource in its inventory of the world's recoverable oil sources in 2004 (Humphries, 2008). On the one hand, Alberta's reserves of bitumen are estimated at more than 1.7 trillion barrels (enough to satisfy the current global demand for over 50 years); and between 2010 and 2035, the project is expected to generate $2,077 billion in new investment, GDP of $2,106 billion, 830,000 new jobs, tax revenue of $311 billion for Canada ($105 billion for Alberta) and more than $623 billion in cumulative royalties for the province, which owns the resource (Honarvar et al., 2011). However, this comes at a significant ecological cost, destroying virgin boreal forest and wildlife habitats; consuming vast quantities of natural gas; removing unsustainable levels of water from the watershed and storing them in lake-sized toxic "tailings" ponds; depleting and poisoning fish stocks; and generating about three times more greenhouse gases per barrel than conventional oil (Nikiforuk, 2010). This is in addition to the immense social and municipal costs, the devastation of traditional Aboriginal lifestyles and the incidence of an unusually high number of cancers downstream of the industry's work (Radford & Thompson, 2011). All of this has tarnished Canada's once-sterling reputation internationally, generating formal, editorial and street protests at home as well as at the highest political levels. Both US President Obama and the European Parliament have addressed groundswell opposition to using "dirty oil" from Canada's bituminous sands (e.g., Pratt, 2010; Sands & Brooymans, 2010; McFarlane, 2009).

In 2008, the Government of Alberta unveiled a three-year, $25-million campaign to create and market a new brand for the province. The government's ensuing research confirmed popular views of Alberta as a place with beautiful scenery, but deficient in its environmental stewardship (Harris/Decima, 2009). The ensuing, new brand – "Freedom To Create. Spirit To Achieve." – included a $4-million online slideshow, *An Open Door* (Alberta, 2009). This production positioned the province as a land of unlimited opportunity and featured pristine landscapes traditionally used to market Alberta as a tourist destination (see Takach, 2013, p. 214, Figure 1).

Theory and Method

In exploring the relationship between the Romantic and extractive gazes in a political context, and in connection with wider, societal processes, this study will examine *An Open Door* through a critical visual discourse analysis.

The concept of discourse has transcended its earlier, speech-based meanings in the areas of conversation analysis and ethnomethodology to embrace the visual. In their discussion of the eminent exponents of critical discourse analysis (CDA), Chouliaraki and Fairclough (1999, p. vii), expressly include "visual images" in their definition of discourse, and CDA is said to have taken a visual turn (Hansen & Machin, 2008, citing Kress & van Leeuwen, 2001, 2006). Theoretical congruence between CDA and visual analysis is suggested further in the two fields respectively studying questions of power and addressing social imbalances in service of seeking positive change. The call for critical attention to prevalent discourses and their attendant power relationships in CDA resonates with calls to challenge naturalized master narratives in visual studies (e.g., Bal, 2003; Pink, 2003; Dikovitskaya, 2005). The ethic of combining CDA and visual analysis is exemplified by Faux and Kim's (2006) study of visual representations of minority victims of Hurricane Katrina, which showed how "different perspectives can inform one another, illustrating how factors such as power and intersecting social forces emerge in situated discourse" (p. 58).

A critical reading of imagery is warranted because images are hardly neutral or objective (Jamieson, 2007; Kress & van Leeuwen, 2001, 2006; Grodal, 2000), being mediated by their creators in their selection and composition, and by viewers in their subjective perceptions. Furthermore, images have social effects (Berger, 1972; Sturken & Cartwright, 2008) and are significant sites to address power struggles in public discourse (Mirzoeff, 2009). As Western society trades increasingly in images and simulacra (Macnaghtan & Urry, 1998; Gierstberg & Osterbaan, 2002), images of nature have been co-opted to serve commercial imperatives, for example, through the "visual consumption" of landscapes solicited by the tourism and entertainment industries (Urry, 1992, pp. 2–3). But not only has such imagery, influenced by the culture of marketing and branding, increasingly abstracted and de-contextualized environments into symbols and icons; it has also "deliberately move[d] away from naturalistic, empirical truth" and thus obscured its connections to flows of power at work in processes like globalization and consumerism (Hansen & Machin, 2008, p. 792).

Our messages around the environment do involve power in society, namely the ability to define our relationship with nature and what we do to it (Hansen, 2010). Diverse constructions and representations of the environment can ideologically validate and advance perspectives in areas such as politics and tourism marketing (Hansen, 2011), and by extension, place branding, which involves distilling and representing the "true brand essence" of a place, i.e. "what it is and what it wants to be known as" (Finucan, 2002, p. 11). Place branding can be used to maintain political and economic hegemony (van Ham, 2010); and as an appeal to collective identity, can obscure lines of power coming from domestic, foreign and transnational sources (Blue, 2008). Thus, examining the branding of a place can help trace those lines of power and reveal the underlying values at work, and specifically its citizens' relationship with their environment, which Corbett (2006) places on a spectrum ranging from purely utilitarian (anthropocentric) through conservationist, preservationist and increasingly ecocentric perspectives.

Visuality can be vital to such investigations. Images are a key element in the formation and negotiation of identities (Emerling, 2008), as well as in our representations of the environment (Hansen, 2010). As Mirzoeff (2009, p. 5) observes, "The disjunctured and fragmented culture that we call postmodern is best imagined and understood visually, just as the nineteenth century was classically represented in the newspaper and the novel." We must ask whether a visual work has "exceeded its essence as the documentation of memory and acquired other values as a symbolic resource for imagination" in constructing a sense of people's heritage (Rusted, 2010, p. 5) or identity. As Remillard (2011, p. 129) explains, "Meaning is not an intrinsic and merely referential property of the image. Rather, visual objects exist as part of larger systems of representations, or discursive formations." At the same time, the modernist notion of reducing the meaning of an image to a single, overarching narrative has given way to multivalent, poststructuralist investigations of systems of power and potential sites for resistance in imagery, as in the work of Tagg (1993, 2009). All of this invites a critical visual discourse analysis here, beginning with some context on Alberta and its tradition of visual representation preceding the government's rebranding campaign.

Alberta and its Prior Visualizations

Alberta forms part of the resource-rich hinterland (relative to more densely populated, easterly parts of North America) extending from Alaska across Western Canada and much of the Western US, and defined by Garreau (1981) as the "Empty Quarter." Since White contact in the eighteenth century, "Alberta" has served diverse purposes. After some ten millennia as Aboriginal territory, the area became a fur farm for the Hudson's Bay Company during the British colonial period. As a territorial district after the Canadian confederation in 1867 and a province since 1905, Alberta's leading industries have included ranching, agriculture and then oil and gas. The major oil strike at Leduc in 1947 swept Alberta from an economic backwater into the modern era (Palmer & Palmer, 1990).

Imagery has played an important part in portraying a sense of identity in the region (Francis, 1992). Reflecting attitudes towards nature prevalent among its European colonizers, the Canadian West has undergone several phases in the popular visual imagination. These are summarized by Francis (1989) as an overlapping evolution from a barren, inhospitable wasteland (from 1650–1850) to both a Romantic, Edenic wilderness and the breadbasket of the British Empire (1845–1885), a utopian promised land for settlers (1880–1920), a harsh place to be conquered in asserting Canadian nationhood (1880–1940), and a mythical "region of the mind" (p. 193) reflecting the views and experiences of its inhabitants.

Western Canada was trumpeted from the 1880s and onward by federal authorities seeking to populate the region as a nation-building exercise, and by the Canadian Pacific Railway for its own commercial purposes, in linking the nation's metropolitan centre to the Pacific coast. The image of the West as a promised land – a vast, unspoiled paradise, typically with endless, windswept fields of grain foregrounding the distant Rocky Mountains, all under an open, azure sky – retains special resonance in Alberta's identity. For many, the province remains "a tabula rasa where people could write their own destinies free of the limitations of privilege and tradition that hampered advancement in the East and in the Old World. [. . .] Even today the rhetoric of the West as Promised Land is used by the Alberta politicians to present the province as a free enterprise conduit and ideological Mecca..." (Francis & Kitzan, 2007, p. xi, p. xxii). The province remains a magnet for economic migration from the rest of Canada to an extent that one sociologist sees as transforming Canadian society (Hiller, 2009). Meanwhile, dominant tourist images of Alberta continue to depict "a wholesome, wild, unsettled cattle country, relatively untouched by the advances of urban and technological culture, and miraculously retaining the spirit of the wild western frontier" (Blue, 2008, p. 75). This is the visual tradition into which the provincial government's rebranding effort was born.

Alberta's Rebranding Slideshow

In analyzing Alberta's rebranding slideshow as visual environmental communication, we can consider three places in which meaning is made in imagery: the site of the slideshow's production – its creators – the site of the image itself and the site of its audience (Rose, 2012). This analysis will focus on what Rose terms the social modality of each site, concerned with the underlying social, economic, political and cultural relations, institutions and identities.

Site of Production

Historically, Alberta's provincial government has enjoyed a privileged position. It has changed only three times since Alberta achieved provincehood in 1905 and only once since 1935. Alberta has generated most of Canada's larger, populist opposition movements over the last century. Massive majority governments are common. These developments have been attributed to factors such as Alberta's distance from the metropolitan centres of political and economic power in central Canada (although the balance of power may be changing with the rise of bituminous-sands extraction); a history of electing federal opposition members in greater numbers and for longer periods than any other Canadian province; and recurring constitutional and other battles with the federal government over the ownership, control and marketing of natural resources found in the province (Melnyk, 1992, 1993; Takach, 2010).

The province has an enduring and deeply engrained tradition of protest, born of a defensive posture in the face of a distant federal authority perceived to be unresponsive to Western concerns. Added to this is a proud commitment to "free enterprise," ardently opposed to restraint by government and manifested in Alberta's proudly boasting the lowest taxes in Canada by far, and being the only province without its own sales tax (Alberta 2012c), even when faced with deficit budgets. This is the political context from which Alberta's rebranding effort grew, one familiar to resource-rich hinterlands vis-à-vis the centres of political and economic power to which they are subservient, and certainly endemic to the northwestern flank of North America, Garreau's (1981) so-called "Empty Quarter." For example, significant political, economic and cultural comparisons have been drawn between Alberta and the American state of Idaho (Alm & Taylor, 2003).

The slideshow was produced by the Alberta government's Public Affairs Bureau (PAB), working with public-relations professionals (Pratt, 2009), and released on the province's official website. The PAB's mission is "to help the government communicate effectively with Albertans and its employees by providing quality, coordinated and cost-effective communications services" (Alberta, 2012b). It has been criticized as a propaganda vehicle, and the province has moved to depoliticize the organization (Kleiss, 2012). The premier's office and other sources confirmed that the slideshow was conceived not just for the customary reasons associated with governmental branding exercises – to attract skilled workers, investment and tourism to the province and to market local products and services – but to counter perceptions of Alberta as a producer of "dirty" oil (Markusoff, 2008; Audette & Henton, 2009). These perceptions stem from not only growing domestic and international protests about the environmental impact of the bituminous-sands project, but also exposés of the government's management of the resource, founded on claims of political corruption and greed (Marsden, 2007) and ecological disaster (Nikiforuk, 2010), and popularized in a *National Geographic* cover story subtitled "Scraping Bottom" (Kunzig, 2009). The province's poor environmental reputation is reflected in repeated failing grades and the lowest national rankings on biodiversity (Sierra Club of Canada, 2006) and climate change (Sierra Club of Canada, 2008), and in well-publicized environmental campaigns to stop the bituminous-sands project (e.g., Corporate Ethics International, 2010; Greenpeace Canada, 2012; UK Tar Sands Network, 2012).

Growing ecological concerns by potential importers of Alberta's synthetic crude are worrisome for the province, given that energy products comprise most of its exports – 71% of them in 2010 (Alberta, 2011a) – and that royalties from the extraction of non-renewable energy resources – primarily the bituminous sands – comprise 27.8 percent of its annual revenue (Alberta, 2012c). The high stakes backgrounding Alberta's slide-show were acknowledged by the then-premier in justifying the $25-million rebranding campaign against an anticipated return of $40 billion in economic benefits to the province (Audette & Henton, 2009). Alberta's position was epitomized by his adamant refusal to "touch the brake" on the accelerated development of the bituminous sands despite pleas from not only a federal government department (Environment Canada), Aboriginal groups, environmental groups, and opposition parties in the Legislature (D'Aliesio & Markusoff, 2008), but also an industrial consortium, though the latter's concerns sprang from cost overruns and labour shortages rather than ecological considerations (Gillespie, 2008). That well-publicized refusal earned the then-premier the title of "Canadian Fossil Fool of the Year" and second place overall in an international competition organized by a coalition of 46 environmental groups, based on an online poll which attracted 6,000 votes (Stevenson, 2008).

Thus, factors at the site of the slideshow's production, which privilege economic considerations (and the progress meta-narrative motivating them), suggest that the intent of the slideshow's producers was more anthropocentric than ecocentric.

The Images

In examining the images of Alberta captured in the slideshow, we may attend to its visual structure and the effects of the images, including content, colour and spatial organization, as well as the organization of the viewer's gaze (Rose, 2012). We could also examine its mix of visuals, sound, language and music (Machin, 2007). The slideshow unfolds in just under 2½ minutes over a sequence of 36 frames or still shots, 27 of which contain photographic images and text, and the remaining 9 of which contain text only. Each shot is accompanied by the voice of an unseen male narrator. No other audio adorns the piece until orchestral music slowly fades into the final half-dozen shots, building to a crescendo at the end. This review will reference specific, exemplary frames of the slideshow, noting accompanying text and music where illustrative.

The primary theme is established in the slideshow's title shot as the words "An open door" appear in mauve and white text on a background of gold. The idea of Alberta as open geographically is illustrated in words appearing onscreen, in identical narration (voiced over those words) and in accompanying imagery, in nine shots. An open door – literally and figuratively a human construction – suggests crossing a threshold to access what lies beyond it. Here the viewer's gaze seems to be that of an explorer or prospector; our invited entrance is not passive or contemplative, but active. We may appreciate and even revere the scenery, but by presenting the landscape as available to us (and perhaps even there *for* us), the show invites direct engagement with Alberta.

The slideshow brims with panoramas of majestic scenery, such as a forest-surrounded Rocky-Mountain Lake (see Takach, 2013, p. 219, Figure 2), a vibrant-yellow canola field and the eerie 'badlands' which contain the world's richest bed of dinosaur fossils. Of the slideshow's 36 shots, 17 feature wide-open spaces, typically under an expanse of azure sky.

Here the slideshow echoes traditional marketing imagery of the province, as well as popular perceptions documented in the government's branding research, of Alberta's landscapes as unspoiled and visually stunning. However, from a place-branding perspective, we should note that Alberta's demographics are overwhelmingly urban. Between 1911 and 1941, the urban population barely moved from 37% to 39%, but following the discovery at Leduc in 1947, it leaped to 73% by 1971 and then to 82% by 2006, ranking Alberta among Canada's most urbanized provinces (Statistics Canada, 2009a; 2009b). Yet the face and the branding image of Alberta remain resolutely pastoral. In the provincial election in 2012, the one-third of Albertans who lived outside of the province's two largest cities, Calgary and Edmonton, were allocated more than half of the seats in the legislature. In popular culture, both images and words continue to glorify a mythical, rural West, largely devoid of people and dominated by vacant landscapes (Melnyk, 1999; Calder, 2001). A typical treatment is a picture-book issued during the provincial centennial, *A is Alberta: A Centennial Alphabet* (Tingley & McInnis, 2005), which allows only 2 cityscapes and 9 images of people (3 cowboys, 2 painters, a canoeist and some tourists) onto its 26 pages.

New-Western historians suggest that this latter-day notion of a mythical West remains a repository of a national frontier past and a site for people to locate a core of their identity. As Cronon, Miles and Gitlin (1992, p. 26) observe of the region, "Rather than a landscape of boundless freedom, it is a walled-off preserve in which the very act of experiencing the wild proves how tame it has become." In keeping with this ethos, only 6 of the 36 shots in Alberta's rebranding slideshow offer visibly urban views, depicting the downtown skylines of Edmonton and Calgary, a wide-angle view of a suburban backdrop and (fleetingly) a night view of an office tower.

Thus, the slideshow offers images of Alberta's picturesque landscapes, but relatively little indication of environments in which Albertans actually live or work. There is no visible representation of geography or communities outside of the province's national or provincial parks or its two large cities. In my reading, when people appear in landscapes, they tend to be positioned as tourists or spectators, for example, standing in a field of wheat, facing a mountain peak or staring out onto a lake, always dwarfed by the vastness of the natural landscape (Takach, 2013, p. 219, Figure 2).

This engages an important issue in environmental communication. For Cronon (1995), the environment is not "a pristine sanctuary" devoid of "the contaminating taint of civilization" but "a product of that civilization" (p. 69). If we adopt a social constructionist (as opposed to an "objective" realist) view of nature, then there is no single "nature," but a diversity of them, each constituted (and contested) through diverse socio-cultural processes – reflecting or opposing dominant ideas of society – with which those natures are inextricably intertwined (Macnaghten & Urry, 1998). As Cronon (1995, p. 7) notes:

> Wilderness hides its unnaturalness behind a mask that is all the more beguiling because it seems so natural. As we gaze into the mirror it holds up for us, we too easily imagine that what we behold is Nature when in fact we see the reflection of our own unexamined longings and desires.

Although sympathetic to ecological concerns, Cronon's influential critique of the longstanding separation of people from nature has been accused of providing an argument *against* environmentalism, as that separation has traditionally grounded Romantic ideas of preserving nature (Saltmarsh, 1995). However, Cronon's argument is that such segregation further disengages us from taking responsibility for nature. The imagery in Alberta's slideshow invites viewers to admire the scenery, whether by way of a tourist gaze (Urry, 1992), a possessory gaze or simply preservationist reverence (Tompkins, 1992; Ingram, 2000). When people in the slideshow interact with nature in a more active way, the context seems consumptive. Images of hikers, a skater and a canoeist enjoying mountain panoramas (in five shots, including the final one, Figure 1), accompanied by textual and narrated references to the province offering "an open door" and an opportunity to "dream big," represent nature as a metaphorical pathway to freedom and personal achievement. We may read this literally in one low-angle shot, which directs our gaze from the ground upwards, to a bare human leg stepping on a fallen tree's skeleton to cross an alpine stream (see Takach, 2013, p. 221, Figure 3): nature bends (and in this case, falls) to permit human passage and use.

Advertising, which asserts a significant and deeply entrenched presence in industrial society, has revised and co-opted environmental meanings and imagery in a proliferation of "green" marketing since the late 1980s (Hansen, 2010). This marks both the "totalizing incorporation of Nature by industrialized culture and the commodification of the Unconscious into a visible in the marketing of 'desire' produced as media" (Brereton, 2005, p. 185). For Hansen (2010), the seamless, almost invisible use of nature in advertising and the innate values it expresses give nature its ideological power. Greenwashing has become a common public-relations strategy to mislead the public as to the ecological soundness of practices or products (Cox, 2013). Like advertising generally, tourism and place branding have co-opted nature in service of consumption. Intrinsically anthropocentric, tourist discourse pitches human comfort, discovery and pleasure. As Todd (2010, p. 220) observes, photographs in particular can "manipulate nature in ways that justify visually consumptive tourist practices and exacerbate the hegemonic relationship between tourist and places."

The images in Alberta's slideshow tend to position landscapes as sources of human opportunity, inspiration and/or pleasure, but the presentation sidesteps notions of the land as providing the economic foundation of the province. Aboriginal people relied on local bison for millennia; after White contact, people in "Alberta" moved primarily from the fur trade into ranching and growing grain until the oil strike at Leduc. Albertans have generated nation-leading wealth over the past two decades (Alberta, 2011b), due significantly to the extraction of non-renewable resources – conventional oil, natural gas, coal, coalbed methane, and oil from the bituminous sands – along with forestry and agriculture. Tourism and other secondary and tertiary industries round out the economy, but the dominance of oil and gas is evidenced by the province's significant reliance on revenue from energy royalties to meet its budgetary expenses (Alberta, 2012c) and the economic uncertainty attendant on volatile commodity prices beyond provincial control. The slideshow's only indication that Alberta produces fossil fuels is in a small, backlit image of an oil pumpjack at sunset, one of 16 photos in a fleeting collage that constitutes one of the last shots. (That collage also includes a shot of a series of wind turbines foregrounding a grain field and the distant Rocky Mountains.) This visual imbalance downplays the extractive gaze in favour of the Romantic one.

The disconnect between people and their environments is echoed in the only image visibly set in a community in which Albertans actually live: two youthful skaters, partly decapitated photographically, foreground an array of huge houses with no other residents in sight. An image of a father holding his young son's hand, their backs to the viewer as they regard a lake in autumn, accompanied by narration and onscreen text ("Albertans care. About the future, not just our own, but of the world."), is the sole explicit reference to environmentalism (see Takach, 2013, p. 222, Figure 4). However, it comes almost two-thirds through the slideshow; does not depict its subjects engaging actively in caring for the wider world; and may be read to suggest caring for our children at least as much as preserving nature – and even then, in the context of the rest of the imagery, perhaps more for (anthropocentric) use by future generations than for ecocentric reasons. These factors tend to diminish the ecocentric impact of the image of the father and son at the lake, and may further blur the distinction between Romantic and extractive gazes.

An ecocentric argument might have been bolstered by including images of the wildlife historically used to depict the province as unspoiled, such as the bighorn sheep and the Great Horned Owl, Alberta's official mammal and bird, respectively. However, the only non-human life appearing in the show are two backlit saddle horses carrying their riders under a limitless, azure sky – another example of nature serving people.

For the government, presenting Alberta as pristine nature in response to international and domestic criticisms of its ecological stewardship around the bituminous sands evidently trumped remedying perceived shortfalls in Albertans' hospitality and compassion, also identified in its pre-branding research. This choice may seem to undercut a primary part of the slideshow's mission, but it may be explained by the political and economic context highlighted above and by turning to the slideshow's audiencing.

Site of Audience

Although the Alberta government's slideshow never identified its target audience directly, the premier's director of communications told the media that the rebranding campaign aimed "to tell the world that we're producing clean energy," adding, "I don't think we can leave that job to Greenpeace and the Sierra Club" (Markusoff, 2008, p. B8). Given the premier's casting of the campaign as an economic investment, it seems plausible that an important, if not primary, target audience for the rebranding campaign and slideshow lay due south in the United States.

Alberta's ties to the US are significant. Long known as "the most Americanized part of Canada" (Simpson, 2000, p. 19), Alberta enjoyed two great waves of American immigration, the first following the evaporation of free land for settlers in the US by the early twentieth century, and the second following the discovery of oil at Leduc, accompanied by substantial American investment and technology. After an early tradition of political cooperativism and radicalism inherited from incoming American farmers, the coming of the modern oil industry in 1947 saw the province adopt a proudly pro-business persona and discourse, even as it maintained a large government apparatus and huge subsidies to business, particularly the oil industry (Richards & Pratt, 1979; Taft, 1997; Takach, 2010).

In addition to its American connections through the energy industry, Alberta has a substantial economic dependency on the US, which buys almost 90% of the province's exports (Alberta, 2011a), while providing two-thirds of Alberta's foreign investment and 60% of its foreign tourists (Alberta, 2011c). These ties were emphasized in recent debates over extending the environmentally controversial Keystone XL pipeline from Alberta's "dirty" bituminous sands to the Gulf of Mexico (Payton, 2012). The US consumes one-quarter of the world's oil production, and boasts only 2% of its reserves (Broder, 2011). Even with increasing domestic shale-oil production, the US looks increasingly to Alberta as a secure alternative to volatile sources such as OPEC nations. As one Congressional report outlines, the advanced state of infrastructure and investment in Alberta's bituminous sands, as well as the ecological damage involved, may make that project a preferable investment to any efforts to recover "unconventional" oil in the US (Humphries, 2008).

The Alberta government's American audiencing seems to have been born out in its announcing the release of the slideshow just a week before President Obama's first scheduled visit to Canada (Pratt, 2009); in purchasing ads in American publications defending the bituminous sands, including one for $55,800 in *The Washington Post*, which that newspaper had rejected as a proposed opinion column (O'Donnell, 2010); in including Americans in its focus testing for the brand (Harris/Decima, 2009); and in the then-premier's reference to correcting misconceptions about Alberta being spread by ecological groups to "Americans and other customers" (Audette & Henton,

2009). In the slideshow itself, 17 textual and narrated references to "Alberta" and "Albertan" – arguably rendered jingoistic by the force of repetition – and an anthemic conclusion that "We hold true to the belief that our path from the past to the future can be made wide enough to carry the dreams of all Alberta ... of all Albertans" seem discernibly American in rhetorical style.

The reception of the slideshow became a public-relations nightmare for the government and its public-relations counsel on the project. Shortly after the show's online launch, a viewer inquired about the location of a featured image of two children frolicking on a beach, reproduced in both the first version of the slideshow and on banner ads. When it emerged that the photograph came from northeastern England and the premier's office defended the selection by claiming it symbolized Albertans' worldliness (Stokes, 2009), the rebranding was mocked at home and abroad (e.g., Wainwright, 2009). Beyond favourable comments from the premier (Audette & Henton, 2009), the slideshow generated little positive response in the popular media. Pundits called the rebranding campaign's overarching slogan clichéd, forgettable and embarrassing (Braid, 2011). Political critics on all sides were similarly pointed. The day's provincial opposition leader, commonly considered to be on the political left of the governing party, declared the slideshow's content irrelevant to most Albertans' lives (Audette & Henton, 2009). From what is widely perceived to be the political right, the Canadian Taxpayers Federation's director stated, "Don't try to pretend that we don't have the oil sands and we simply are just a bunch of parks and people who are out there to achieve and create or whatever the words were" (Canadian Press, 2009).

As of November 5, 2011, the slideshow had received 4,716 hits on YouTube compared to 97,213 hits of an online video subsequently produced by a US-based group (Corporate Ethics International, 2010) urging viewers to *boycott* Alberta because of the environmental devastation wrought by the bituminous sands. When a new premier took office upon her predecessor's retirement in late 2011, she found the rebranding in disuse and in need of phasing out (Braid, 2011), and the slideshow was removed from the government's web page.

Thus, the reception of the slideshow reinforces the pre-eminence of economic (and anthropomorphic) considerations, despite the prevalence of pristine images of nature that may, at least initially, seem to be more ecocentric and inviting a Romantic gaze. This leads to concluding notes on the broader societal processes at work beneath the picturesque imagery.

Societal Processes, Anthropocentricism and Extractive/Romantic Gazes

Alberta's laissez-faire attitude to economic growth – which critics declare to be at the unsustainable expense of ecological concerns – fits clearly into the neoliberal model, privileging the profit motive as the primary basis for organizing human affairs and characterized by government policies such as lower public spending, deregulation and reduced barriers to trade and foreign investment (Couldry, 2010). Although neoliberalism has held sway on the global scene since the 1980s, its limits and flaws are evidenced by, for example, the widening gap between the world's rich and poor

(Primorac, 2011), the economic collapse of 2008 and the Occupy movement of 2011. Neoliberal marketing strategies have seamlessly co-opted environmental discourse in everything from advertising consumer products, services and tourism destinations to branding places and their identities, all trading on deep-rooted, popular and positive associations with nature and "the natural" (Hansen, 2010).

As the Alberta government's rebranding slideshow, *An Open Door*, tells us, extractive capitalism co-opts landscapes not only in industries like tourism, but also in broader political and economic processes of self-definition and marketing through place branding. Viewing the slideshow not only through its imagery but also through the eyes of its producers and its audience leads to three conclusions for the emerging discipline of visual environmental communication.

First, on Corbett's (2006) environmental spectrum, the slideshow's themes of opportunity, freedom and personal achievement – juxtaposed against a preponderance of images of pristine, majestic nature – may be seen as an ecocentric, Romantic gloss on an anthropocentric core, rooted firmly in a consumptive gaze explicable as both extractive and Romantic. Nature, beautiful as it is, is presented as available for human conquest and advantage, a grand backdrop for the economic Darwinism valorized by neoliberal forces under the mantles of globalization and free enterprise. Ironically, the ultimate beneficiaries of such picturesque place branding may be the investors and corporate executives profiting from exploiting and destroying pristine landscapes like those depicted in that branding. Often, these people live far beyond the places that their companies exploit: environmental sacrifice zones like Alberta's bituminous sands. Thus, Corbett's taxonomy can be applied not only to the content of visual communication, but also to its context, which, as this study suggests, may fall on a different spot on the spectrum of anthropocentric and ecocentric interests than the content might suggest.

Second, this case study provides further support for the argument linking not only the Romantic and extractive gazes to each other in contemporary visual studies (Hodgins & Thompson, 2011), but also potentially linking such a combined, consumptive gaze to broader social processes such as globalization, neoliberalism and Western-style consumerism. This follows and supports the investigations of Cronon (1995) and Morton (2007) into the misleading separation of people from nature, and the argument of Hansen and Machin (2008) for looking past the iconicity of imagery – sometimes so deceptively simple – to reveal fuller understandings of the societal forces and power relations involved in its deployment. In that light, the link between the seemingly benign Romantic gaze (proffered by proponents of economic development and environmental protection alike for their respective purposes) and the extractive gaze (proffered more dramatically by ecocentric interests and more subtly by pro-development forces) may seem stronger than their apparent inconsistency might have suggested in less ecologically conscious times.

Finally and more broadly, Alberta's rebranding effort reminds us that postmodern society can be considered ocularcentric "not simply because visual images are more and more common, nor because knowledges about the world are increasingly articulated visually, but because we interact increasingly with completely constructed visual experiences" (Rose, 2012, p. 4, citing Mirzoeff, 1998). In environmental communication, this supports Cronon's (1995) notion of nature as a subjective construction, an apparently neutral image reflecting the unexpressed desires of the people gazing at the image. But perhaps more than anything else, a rebranding exercise like *An Open Door* also speaks to the inherent ability – and great power – of images to depict one thing and mean not only something different, but substantially more than what they depict. In this sense, given the escalating stakes in the global conflict between economic development based on extractive capitalism and its increasingly unsustainable environmental costs, visuality brings a responsibility as formidable as its power.

Acknowledgements

The author heartily thanks the editors and reviewers for their wise, generous and helpful advice, and the munificent funders of this research, the Social Sciences and Humanities Research Council of Canada and the University of Calgary.

References

Alberta (Government). (2009). Showcasing Alberta: Experience the Alberta story: An open door. Retrieved November 4, 2011 from http://albertan.ca/showcasingalberta/158.htm
Alberta (Government). (2011a). Alberta international exports 2010. Retrieved July 12, 2012 from http://www.international.alberta.ca/documents/Alberta_International_Exports_2010.pdf
Alberta (Government). (2011b). Highlights of the Alberta economy 2011. Retrieved July 12, 2012 from http://www.albertacanada.com/documents/SP-EH_highlightsABEconomy.pdf
Alberta (Government). (2011c). US–Alberta relations. Retrieved July 12, 2012 from http://www.international.alberta.ca/documents/International/US-AB.pdf
Alberta (Government). (2012a.) Energy: Facts and statistics. Retrieved July 6, 2012 from http://www.energy.gov.ab.ca/OilSands/791.asp
Alberta (Government). (2012b). Executive Council: Public Affairs Bureau: About us. Retrieved July 11, 2012 from http://publicaffairs.alberta.ca/520.cfm
Alberta (Government). (2012c). Overview: Budget 2012: Investing in people. Retrieved June 24, 2012 from http://budget2012.alberta.ca/fact-card-economic-revenue-highlights.pdf
Alm, L.R., & Taylor, L. (2003). Alberta and Idaho: An implicit bond. *American Review of Canadian Studies, 33*(2), 197–218.
Audette, T., & Henton, D. (2009). New brand to counteract image of "dirty oil," Province to spend $25M over three years. *Edmonton Journal*, March 16, A1.
Bal, M. (2003). Visual essentialism and the object of visual culture. *Journal of Visual Culture, 2*(1), 5–32.
Berger, J. (1972). *Ways of seeing*. London: BBC and Hammondsworth.
Blue, G. (2008). If it ain't Alberta, it ain't beef: Local food, regional identity, (inter)national politics. *Food, Culture and Society, 11*(1), 69–85.
Braid, D. (2011). "Spirit to Achieve" headed for exits. *Calgary Herald*, October 26, A1.
Brereton, P. (2005). *Hollywood utopia: Ecology in contemporary American cinema*. Bristol, UK: Intellect.
Broder, J.M. (2011). Obama lays out plan to cut reliance on fuel imports. *New York Times*, March 31: NA(L).
Calder, A. (2001). Who's from the prairie? Some prairie self-representations in popular culture. In R. Wardhaugh (Ed.), *Toward defining the prairies: Region, culture, history* (pp. 91–100). Winnipeg, MB: University of Manitoba Press.

Cammaer, G. (2009). Edward Burtynsky's *Manufactured Landscapes:* The ethics and aesthetics of creating moving still images and stilling moving images of ecological disasters. *Environmental Communication, 3*(1), 121–130.

Canadian Press. (2009). Alberta unveils new logo: "Freedom to create, spirit to achieve." *Trail Times* [British Columbia], March 26, p. 5.

Chastko, P. (2004). *Developing Alberta's oil sands: From Karl Clark to Kyoto.* Calgary, AB: University of Calgary Press.

Chouliaraki, L., & Fairclough, N. (1999). *Discourse in late modernity.* Edinburgh, UK: University of Edinburgh Press.

Clarke, T. (2008). *Tar sands showdown: Canada and the new politics of oil in an age of climate change.* Toronto, ON: Lorimer & Co.

Corbett, J. (2006). *Communicating nature: How we create and understand environmental messages.* Washington, DC: Island Press.

Corporate Ethics International. (2010). Rethink Alberta. Retrieved June 24, 2012 from http://www.rethinkalberta.com/press.php?r=4

Couldry, N. (2010). *Why voice matters: Culture and politics after neoliberalism.* London: Sage.

Cox, R. (2013). *Environmental communication and the public sphere.* 2nd ed. Thousand Oaks, CA: Sage.

Cronon, W. (1995). The trouble with wilderness; or, getting back to the wrong nature. In W. Cronon (Ed.), *Uncommon ground: Toward reinventing nature* (pp. 69–90). New York: W.W. Norton.

Cronon, W., Miles, G., & Gitlin, J. (Eds.). (1992). *Becoming West: Toward a new meaning for Western history.* New York: W.W. Norton.

D'Aliesio, R., & Markusoff, J. (2008). No brakes on oilsands: Stelmach won't suspend land leases for now. *Edmonton Journal,* February 26, p. A1.

Dikovitskaya, M. (2005). A look at visual studies. *Afterimage, 29*(5), 4.

Emerling, J. (2008). Jacques Ranciere, the future of the image. London: Verso, 2007. *Journal of Visual Culture, 7*(3), 376–381.

Faux, W.V., & Kim, H. (2006). Visual representation of the victims of Hurricane Katrina: A dialectical approach to content analysis and discourse. *Space and Culture 9,* 55–60.

Finucan, K. (2002). What brand are you? *Planning, 68,* 10–13.

Francis, R.D. (1989). *Images of the West: Changing perceptions of the prairies, 1690–1960.* Saskatoon, SK: Western Producer Prairie Books.

Francis, R.D. (1992). In search of prairie myth: A survey of the intellectual and cultural historiography of prairie Canada. *Journal of Canadian Studies, 24*(3), 44–69.

Francis, R.D., & Kitzan, C. (Eds.). (2007). *The prairie West as promised land.* Calgary, AB: University of Calgary Press.

Garreau, J. (1981). *The nine nations of North America.* Boston, MA: Houghton Mifflin.

Gierstberg, F.O., & Osterbaan, W. (Eds.). (2002). *The image society: Essays on visual culture.* Rotterdam, Netherlands: NAi Publishers.

Gillespie, C. (2008). Scar sands. *Canadian Geographic, 128*(3), 64–78.

Greenpeace Canada. (2012). Our campaigns: Tar sands. Retrieved July 12, 2012 from http://www.greenpeace.org/canada/en/campaigns/Energy/tarsands

Grodal, T.K. (2000). Subjectivity, objectivity and aesthetic feelings in film. In Ib Bondenbjerg (Ed.), *Moving images, culture and the mind* (pp. 87–104). Luton, UK: University of Luton Press.

Hansen, A. (2010). *Environment, media and communication.* Oxon, UK: Routledge.

Hansen, A. (2011). Communication, media and environment: Towards reconnecting research on the production, content and social implications of environmental communication. *International Communication Gazette, 73*(1–2), 7–25.

Hansen, A., & Machin, D. (2008). Visually branding the environment: Climate change as a marketing opportunity. *Discourse Studies, 10*(6), 777–794.

Harris/Decima. (2009). *Research summary: Branding Alberta initiative.* Retrieved July 12, 2012 from http://alberta.ca/albertacode/documents/2009BrandResearchSummary.pdf

Hiller, H.H. (2009). *Second promised land: Migration to Alberta and the transformation of Canadian society*. Montreal, QC/Kingston, ON: McGill-Queen's University Press.

Hodgins, P., & Thompson, P. (2011). Taking the romance out of extraction: Contemporary Canadian artists and the subversion of the romantic/extractive gaze. *Environmental Communication, 5*(4), 393–410.

Honarvar, A., et al. (2011). *Economic impacts of new oil sands projects in Alberta (2010–2035)*. Calgary, AB: Canadian Energy Research Institute/University of Calgary.

Humphries, M. (2008). *CRS [Congressional Research Service] report for Congress: North American oil sands: History of development, prospects for the future*. Retrieved July 12, 2012 from http://www.fas.org/sgp/crs/misc/RL34258.pdf

Ingram, D. (2000). *Green screen: Environmentalism and Hollywood cinema*. Exeter, UK: University of Exeter Press.

Jamieson, H. (2007). *Visual communication: More than meets the eye*. Bristol, UK: Intellect Books.

Kleiss, K. (2012). Redford getting party word out. *Edmonton Journal*, June 9, p. A1.

Kress, G., & van Leeuwen, T. (2001). *Multimodal discourse: The modes and media of contemporary communication*. London: Arnold.

Kress, G., & van Leeuwen, T. (2006). *Reading images: The grammar of visual design*. 2nd ed. London: Routledge.

Kunzig, R. (2009). The Canadian oil boom: Scraping bottom. *National Geographic, 215*(3), 34–59.

Leahy, S. (2006). Burning energy to produce it. Retrieved July 12, 2012 from http://www.ipsnews.net/news.asp?idnews=34076

Machin, D. (2007). *Introduction to multimodal analysis*. London: Hodder Arnold.

Macnaghten, P., & Urry, J. (1998). *Contested natures*. London: Sage.

Markusoff, J. (2008). Alberta's image getting makeover to battle its environmental rep. *Edmonton Journal*, April 24, p. B8.

Marsden, W. (2007). *Stupid to the last drop: How Alberta is bringing environmental Armageddon to Canada (and doesn't seem to care)*. Toronto, ON: Knopf.

McFarlane, A. (2009). Camp targets BP oil plan. *BBC News*, September 1. Retrieved July 12, 2012 from http://news.bbc.co.uk/2/hi/uk_news/8232522.stm

Melnyk, G. (1992). *Riel to Reform: A history of protest in Western Canada*. Saskatoon, SK: Fifth House.

Melnyk, G. (1993). *Beyond alienation: Political essays on the West*. Calgary, AB: Detselig.

Melnyk, G. (1999). *New moon at Batoche: Reflections on the urban prairie*. Banff, AB: Banff Centre Press.

Mirzoeff. N. (1998). *An introduction to visual culture*. 1st ed. London: Routledge.

Mirzoeff. N. (2009). *An introduction to visual culture*. 2nd ed. London: Routledge.

Morton, T. (2007). *Ecology without nature: Rethinking environmental aesthetics*. Boston, MA: Harvard University Press.

Nikiforuk, A. (2010). *Tar sands: Dirty oil and the future of a continent* (Rev. ed.). Vancouver, BC: Greystone.

O'Donnell, S. (2010). Alberta pays to deliver oilsands message. *Edmonton Journal*, July 3, p, A1.

Palmer, H., & Palmer, T. (1990). *Alberta: A new history*. Edmonton, AB: Hurtig.

Payton, L. (2012). Keystone XL pipeline proposal rejected – for now. *CBC News*, January 19. Retrieved July 12, 2012 from http://www.cbc.ca/news/politics/story/2012/01/18/pol-keystone-xl-pipeline.html

Pink, S. (2003). Interdisciplinary agendas in visual research: Re-situating visual anthropology. *Visual Studies, 18*(2), 179–193.

Pratt, S. (2009). Tories drop iconic slogan. *Calgary Herald*, February 24, p. A1.

Pratt, S. (2010). Eyes of the world are watching Alberta. *Edmonton Journal*, October 3, p. A1.

Primorac, M. (2011). F&D spotlights widening gap between rich and poor. Retrieved July 12, 2012 from http://www.imf.org/external/pubs/ft/survey/so/2011/new091211a.htm

Radford, T., & Thompson, N. [dirs.] (2011). *Tipping point: The age of the tar sands* [documentary film]. Broadcast on CBC Television, January 27. Retrieved June 23, 2012 from http://www.cbc.ca/documentaries/natureofthings/video.html?ID=1769597772

Remillard, C. (2011). Picturing environmental risk: The Canadian oil sands and the National Geographic. *International Communication Gazette, 73*(1–2), 127–143.

Richards, J., & Pratt, L. (1979). *Prairie capitalism: Power and influence in the new West.* Toronto, ON: McClelland & Stewart.

Rose, G. (2012). *Visual methodologies: An introduction to the interpretation of visual materials.* 3rd ed. London: Sage.

Rusted, B. (Ed.). (2010). *The art of the Calgary Stampede.* Calgary, AB: Nickle Arts Museum.

Saltmarsh, J.A. (1996). *Uncommon ground: Toward reinventing nature.* Edited by William Cronon [book review]. *New England Quarterly 69*(4), 680–682.

Sands, A., & Brooymans, H. (2010). Rough welcome for Alberta minister at Mexico climate talks. *Postmedia News*, December 6. Retrieved July 12, 2012 from http://www.edmontonjournal.com/business/Rough+welcome+Alberta+minister+Mexico+climate+talks/3932124/story.html

Sierra Club of Canada. (2006). *Rio Report.* Retrieved July 12, 2012 from http://www.sierraclub.ca/national/rio [individual annual reports available by year]

Sierra Club of Canada. (2008). *Kyoto Report Card.* Retrieved July 12, 2012 from http://www.sierraclub.ca/national/kyoto/index.html [individual annual reports available by year]

Simpson, J. (2000). *Star-spangled Canadians: Canadians living the American dream.* Toronto, ON: Harper Collins.

Statistics Canada. (2009a). Population urban and rural, by province and territory (Alberta). Retrieved July 12, 2012 from http://www.statcan.gc.ca/tables-tableaux/sum-som/l01/cst01/demo62j-eng.htm

Statistics Canada. (2009b). Population urban and rural, by province and territory (Canada). Retrieved July 12, 2012 from http://www.statcan.gc.ca/tables-tableaux/sum-som/l01/cst01/demo62a-eng.htm

Stevenson, J. (2008). U.S. environmental coalition names Alta premier as runner-up Fossil Fool. Canadian Press, March 31. Retrieved July 12, 2012 from http://search.proquest.com.ezproxy.ae.talonline.ca/docview/360089014?accountid=46585

Stokes, P. (2009). Canadian province uses Northumberland Beach images to promote tourism. *Telegraph* [London, UK], April 24. Retrieved July 12, 2012 from http://www.telegraph.co.uk/news/newstopics/howaboutthat/5213125/Canadian-province-uses-Northumberland-beach-images-to-promote-tourism.html

Sturken, M., & Cartwright, L. (2008). *Practices of looking: An introduction to visual culture.* 2nd ed. Oxford: Oxford University Press.

Tagg, J. (1993). *The burden of representation: Essays on photographies and histories.* Minnesota, MN: University of Minnesota Press.

Tagg, J. (2009). *The disciplinary frame: Photographic truths and the capture of meaning.* Minneapolis, MN: University of Minnesota Press.

Taft, K. (1997). *Shredding the public interest: Ralph Klein and 25 years of one-party government.* Edmonton, AB: University of Alberta Press/Parkland Institute.

Takach, G. (2010). *Will the real Alberta please stand up?* Edmonton, AB: University of Alberta Press.

Takach, G. (2013). Selling nature in a resource-based economy: Romantic-extractive gazes and Alberta's bituminous sands. *Environmental Communication, (7)*2, 211–230.

Tingley, K., & McInnis, R.F.M. (2005). *A is Alberta: A centennial alphabet.* Regina, SK: Simple Truth Publications.

Todd, A.M. (2010). Anthropocentric distance in *National Geographic*'s environmental aesthetic. *Environmental Communication, 4*(2), 206–224.

Tompkins, J. (1992). *West of everything: The inner life of Westerns.* Oxford: Oxford University Press.

UK Tar Sands Network. 2012. Campaigns. Retrieved June 24, 2012 from http://www.no-tar-sands.org/campaigns
Urry, J. (1992). The tourist gaze and the "environment." *Theory, Culture & Society*, 9(3), 1–26.
van Ham, P. (2010). *Social power in international politics*. Oxon, UK: Routledge.
Wainwright, M. (2009). Canada tourist video shot in Northumbria. *The Guardian*, April 25. Retrieved July 10, 2012 from http://www.guardian.co.uk/world/2009/apr/25/ottawa-northumberland-advert

"Single-minded, compelling, and unique": Visual Communications, Landscape, and the Calculated Aesthetic of Place Branding

Nicole Porter

Place branding strategies play a significant role in the professional composition of landscape imagery, including the depiction of "natural" landscapes. In this paper, Brand Blue Mountains, a brand currently implemented in the Blue Mountains region (Australia), is discursively analyzed. The brand sets out an all-encompassing "Vision" defining the identity, values and personality of the World Heritage listed Blue Mountains landscape, summarized in the tagline Elevate Your Senses. This "vision" is visually translated into a strictly coordinated and copyrighted suite of logos, graphic design, color, fonts and various photographic styles. Analysis reveals that the degree of control that place brand strategists seek to exert over the visual expression of landscape identity is significant. A highly selective narrative of positive nature-based sensory experience is constructed through the holistic application of contemporary visual media. The brands' communications strategy naturalizes and reinforces a particular market-friendly version of place. The framework that brands set for the representation of landscapes overall amounts to an exercise in calculated aesthetics, whereby the form and content of landscape images of various kinds is measured to achieve the greatest market differentiation and impact which technologies allow. The result of this calculated aesthetic system, with its taglines, saturated color, careful composition and magazine-format brevity, is a reduction in the complexity of landscape representations and a perpetuation of nature stereotypes.

Nicole Porter, Ph.D., is faculty member at the University of Nottingham.

Introduction

> BRAND: The totality of images, ideas and reputations of the organization [or place] in the minds of the people who come into contact with it. (Blue Mountains Tourism Limited [BMTL], 2004, p. 46)

Place branding encompasses a range of strategies which deal with the management and marketing of places. These strategies address both visual and non-visual communications in a bid to express consistent, memorable and marketable place identities. As such, place branding strategies play a significant role in the composition of landscape imagery, including "natural" environments. Although the use of nature and landscape imagery in advertising and marketing is not new, the way in which such imagery is systematically coordinated by a number of professionals – brand strategists, tourism managers, graphic designers and photographers – within an overarching brand "totality" puts this imagery in a new context. How do place brand strategies do this, and what are the visual communications that result?

To explore these questions, this paper presents a detailed illustrative case study of one such branded landscape: the Blue Mountains region of New South Wales, Australia. *Brand Blue Mountains* (BBM) was commissioned by local government authority Blue Mountains City Council in 2004, and is still being implemented today. As an area of recognized natural beauty and ecological significance, the Blue Mountains have a long history of visual representation and tourism marketing, of which BBM is the most recent manifestation.

The main BBM document referenced throughout this paper is the Brand Manual (BMTL, 2004), the official in-house guide that defines the brand from its overall "Vision" through to every detail of its expression and management. The Manual sets out key landscape values and motifs and prescribes how these should be communicated through a coordinated suite of logos, graphic design, color, fonts and photography.

The paper is divided into three parts. Firstly, place branding is defined and the key elements of place branding practice, as described within recent literature in the field, are summarized. Secondly, *Brand Blue Mountains* is presented, starting with a background to the brand (context) before presenting a summary of the brand's key characteristics (content). This is followed by a detailed analysis of its visual communications components, before concluding with an account of how these are coordinated and controlled. Finally, the role of visual communications within BBM is critically discussed, highlighting the implications of place branding within discourses of landscape and nature.

Place Branding

Place branding is "the creation of a recognizable place identity and the subsequent use of that identity to further other desirable processes, whether financial investment, changes in user behavior or generating political capital" (Kavaratzis, 2005, p. 334). It is a pervasive, strategic tool used to establish and manage the meaningful sets of relations between things, people, images, texts and physical environments, typically with a view to increasing their market appeal. As an emerging specialization within the branding profession, place branding applies the same kind of techniques used to brand consumer products or companies to the promotion of places.

Branding is characterized by several fundamental strategies, which are applicable whether selling a soft drink, a corporation, a National Park, a city or a nation. The following summary identifies the generic characteristics of the branding process and briefly notes the implications of applying them to places, especially so-called natural landscapes (see also Arvidsson, 2006; Lury, 2004).

First, brands use market research to generate and test marketable narratives. While this may produce more targeted and effective communications, it also means messages may be composed to meet the perceived tastes of selected market demographics at the expense of other possible messages and audiences.

Second, brands are based on projecting uniqueness – identifying or constructing a point of difference with the intention of making one place stand out from others. This has implications for the kinds of landscape narratives and meanings that are conveyed through brands. A place with important ecological or cultural features may be inherently valuable and unique but may not necessarily be brandable; typically this means being unique by being the rarest, largest, oldest, newest, most accessible and so on. According to the literature, in an information-rich world messages must be simplified, straightforward and continually self-reinforcing in order to effectively gain mindshare. Place brand specialist Simon Anholt has described how he has exercised such selectivity in practice:

> On more than one occasion, I have been faced with the tricky task of gently explaining to a very proud and very patriotic minister that the world will not be enthralled by the fact that [...] over sixty species of wild grass grow along his eastern coastline. (2004, pp. 36–7)

Because place branding "provides a base for identifying and uniting a wide range of images [...] *in one marketing message*" (Kavaratzis, 2004, p. 63, emphasis added), a single message cannot pick up on subtle nuances; in such a discourse coastal grasses are not suitable fodder for constructing place image.

Third, branding theory espouses the need for products and companies to express internally consistent values and images. Brands are reductive in their push to reinforce key messages. Anholt suggests "one of the best known functions of brands is to act as convenient, everyday shorthand for what a product or company stands for: why not for a city or country too? Both are handy reductions for far more complex and contradictory realities" (2004, p. 29). Advocates suggest brands make life easier by simplifying the many messages we receive and by introducing "a certain order or coherence to the multiform reality around us" (Mommaas, 2002, p. 34). Critics assert this approach can lead to stereotypical or "simple-minded" place constructs (Gold & Revill, 2004, p. 206).

Fourth, branding discourse asserts that the place identities they construct arise from the qualities of the places they are applied to. Branding consultants regularly describe a process of interpreting existing characteristics of place, defining their task as one of discovering, interpreting, expressing and accentuating an existing identity or essence (see Anholt, 2004, p. 34; Morgan & Pritchard, 2004, p.64). However, such claims are problematic, since to achieve a marketable and coherent identity only *certain* characteristics of place will be included and others excluded.

Fifth, brands are typically emotive and expressive – appealing to the hearts as well as to the minds of consumers. This reflects the "increased focus among marketers on differentiation through relationships and emotional appeals, rather than through discernible, tangible benefits" (Morgan & Pritchard, 2004: p. 61). Brands emphasize qualitative aspects, defining and then continuously attributing particular values (wild, enlivening, exotic, pure, vibrant, authentic and so on) to that which is being branded.

Sixth, brands emphasize interactivity – being a part of something – particularly through the notion of experience (Lindstrom, 2005; Schmitt, 1999). In place branding terms, this manifests itself through a concern with the senses, activities and movement. Notably, branding's concern for emotiveness, values and experiences make landscapes very brandable, as they readily lend themselves to being interpreted as possessing certain qualities (e.g. "natural" or "pure") and they can be directly experienced in space and time (Hankinson, cited in Kavaratzis, 2005, p. 338; Middleton, 2001, p. 355).

Finally, branding strategies are based on the notion of holistic communications, extending beyond conventional advertising to encompass a variety of coordinated communicative acts. Brand narratives are communicated via a range of "surfaces, screens and sites" (Lury, 2004, p. 50): print media surfaces such as those produced by photographers, copyrighters and graphic designers; television and computer screens produced by web designers, cinematographers, directors and so on; and physical sites designed by architects, landscape architects and planners.

Case Study: Brand Blue Mountains

Background

The Blue Mountains region of NSW, located 50 km west of Sydney, consists of over one million hectares of dramatic sandstone escarpments and eucalypt dominated forests of ecological and cultural significance. The Indigenous people of the region – the Gundungurra, Darug and Wiradjuri – have deep associations with this country stretching back over 22,000 years. For colonial settlers this rugged landscape has in turn been perceived as "barren," a "recreational haven" (Hartig, 1987) and most recently as a World Heritage listed wilderness. Unlike other iconic mountain peaks that are typically encountered from below, the Blue "Mountains" – which are actually escarpments – are unique in that they are accessed via ridge tops and plateaus, and are engaged with from above. Inaccessible topography has protected the vast majority of this landscape from permanent human development; however, a corridor of settlements has been established. The City of the Blue Mountains consists of approximately 73,000 residents distributed across 110 km of ridgeline in 27 towns and villages (Blue Mountains City Council [BMCC], 2011). Currently, tourism is the region's primary industry.

The Blue Mountains have a long history of visual representation (Porter & Bull, forthcoming). Throughout the late nineteenth to early twentieth centuries, the Blue Mountains became one of the most important sites where Australians and tourists engaged with the "bush" (woodland), a trend consistent with the rise of mass tourism in other mountain landscapes such as the Lake District in England and Yosemite in the USA. A legacy of accompanying place names, written histories, myths, artworks and advertisements – and especially photographic images – have imbued the Blue Mountains landscape with meaning and contributed to its presence in the cultural imagination. Guidebooks, souvenir books, postcards, maps and newspaper advertisements reflected and influenced the social activities and aesthetic tastes of the day. According to estimates, over 100 different editions of guidebooks, the same number of souvenir books, and somewhere in the order of 5000 postcards have been produced since the 1880s, which when multiplied by the number of print runs amounts to millions of images (Smith, 1998, p. 93). The content of these historic documents vary, sometimes featuring images of the scenery from lookouts and bush walks, and sometimes emphasizing man-made attractions. The formal characteristics of these materials also vary, from elaborate black and white etchings and drawings in the 1880s through to kitsch calendars on sale today.

Constructing BBM

By the twenty-first century, the Blue Mountains were no longer attracting the impressive tourist numbers they had enjoyed a century earlier. In a 2004 study commissioned by the local government tourism body Blue Mountains Tourism Limited (BMTL), a consultant team attributed the decline in visitor interest to a lack of awareness about the kinds of nature experiences the landscape offered, arguing that "the scenery, activities and facilities are as good, if not better than before," but that the perception of the place was the problem, as "potential customers" found the landscape "increasingly unappealing or irrelevant" (Global Tourism & Leisure [GTL], 2004, p. 2). The popularity of the landscape had waned in part due to the earlier success of repeated industry promotions, which had reduced this particular landscape to a well-trodden post-card cliché of scenic panoramas and touristy tea rooms. In response, Blue Mountains City Council commissioned London-based place branding specialists Acanchi to devise BBM, a communications strategy comprehensively outlining how the region could be represented. Sydney-based design firm Infographics developed the visual identity of the brand based on Acanchi's strategic plan.

The strategy was developed over an 18-month period, which included market research and on-site consultation. According to Acanchi:

> Capturing the essence of the Blue Mountains in a single image was a primary goal of the Brand Blue Mountains branding project; an icon which created an immediate resonance with its primary target markets while being single-minded, compelling and unique. (BMTL, 2004, p. 11)

Blue Mountains Tourism Limited (BMTL) owns copyright for the brand and manages its dissemination, most visibly through the official Visit Blue Mountains website (www.visitbluemountains.com.au). To see the brand exclusively as a tourism initiative, however, would be to underestimate its ambition and reach. The brand is intended to not only support tourism and business, but also to function as a "rallying call" for the entire Blue Mountains community to "achieve its economic and social goals" (BMTL, 2004, p. 7). As such, the brand has been implemented by numerous parties, including Blue Mountains City Council, non-government organizations and private businesses who signed up as "brand partners" and thereby agreed to collectively promote the brand message. By 2012, BBM had around 200 partners, ranging from schools and resorts through to local tradespeople and other small businesses (Blue Mountains Business Advantage [BMBA], 2012). In short, the brand's strategic aim is to "unite tourism, business, council and community under a common branding" (BMTL, 2004, p. 5).

The all-encompassing Blue Mountains identity which Acanchi defines for all to follow is "Upness," which it explains as: "A uniquely Australian perspective on being 'UP'; seeing environment, culture, community and commerce as a continuum of experience and engagement. This is a challenge to dated, traditional views of the BM and [sic] to bring understanding to 'a higher level'" (BMTL, 2004, p. 25).

By equating the mountains with upness, Acanchi fashions a connection between the environment's unique defining spatial characteristic – topographical height as experienced from an escarpment – and the activities and sensations it can afford. This relentlessly positive "essence" is then expressed in more detail in the following ways (BMTL, 2004, pp. 6–8):

Brand Positioning: Experience Australia / NSW / Business at another level

Brand Personality: Intelligent and Responsible; Fertile, Imaginative and Vibrant; Unique, Eternal and Wholesome; Earth Centered and One with nature

Brand Values: Integrity and Purity; Creativity; Diversity; Spirit of Achievement; Community Spirit; Sustainability

Tag line (tourism): Elevate Your Senses

Although claiming to be new, this construct perpetuates a well-rehearsed romantic association between mountain landscapes and nature, purity, health, sensory pleasure, timelessness and spiritual rejuvenation and enlightenment (for a full discussion of such associations, see Scharma, 1995).

To communicate these landscape concepts in today's media environment, a 46-page Brand Manual ("the Manual") prescribes a variety of visual and typographic devices to deliver the brand message. The Manual articulates the brand concept before illustrating the visual branding activities to be undertaken by BMTL and brand partners. The Manual "directs how communications messages are crafted, how a visual design is created and how photography is shot, to deliver a consistent and motivating message" (BMTL, 2004, p. 7).

The translation of brand concept to brand image is made explicit through detailed explanations, instructions and examples of the main visual elements to be used. BBM is quick to point out that it is more than just a rubber-stamp image or a single logo, and is based upon an integrated, comprehensive communications strategy "which uses a variety of visual and typographic devices to deliver its message" (BMTL, 2004, p. 11). It explicitly details the content and form of all visual imagery as part of the brand strategy. As a document intended for the use of BMTL and its stakeholders, the Manual is a guide to the internal workings of place branding in relation to landscape concepts and how they are articulated.

Visualizing the Brand: Global Elements (Logo, Colors, Fonts, and Graphic Layout)

BBM sets out to shift public perception from what its creators believe is a "traditional, mono-dimensional view" (BMTL, 2004, p. 25) of the landscape. To achieve this, it employs a communications strategy that it claims differs visually and conceptually from existing marketing and photographic conventions for the region. The brand's graphic design repertoire of signature, typeface, color palette and page layout, known collectively as "global elements," interprets existing landscape characteristics and pictorially represents these in commercialized graphic form.

The most recognizable and regularly applied global element is the "signature," consisting of a graphic element (logo) and a typemark (Figure 1). The logo component of the signature is made up of two sub-graphics that the Manual names "Flora" and "Escarpment" (BMTL, 2004, p. 21), along with a green ribbon-like gestural mark that is reminiscent of a ridge top road. The typemark element consists of the words "blue mountains" carefully composed according to color (blue), font and boldness. The letters comprising the word "blue" are horizontally misaligned to give the impression of a rising and falling horizon. Thus, in its overall composition, the brand signature is a landscape in itself; a miniaturized, abstracted graphic and typographic evocation of form and character.

Although the signature is sufficiently abstract to allow multiple subjective interpretations, it is intended to be read in part as a landscape: "[it] will mean different things to different people [...] Many people see natural elements such as the escarpment, the ridgelines and floral elements. Others see creativity, celebration and our community spirit" (BMTL, 2004, p. 11).

The signature is designed to be either reproduced in full or adapted and applied to suit different publications and contexts. BMTL makes frequent use of part of the signature ("sub-graphics") in their publications, further abstracting the already-abstract landscape motifs into cropped dashes of color. Their web pages animate the floral sub-graphic so that it appears to grow up the screen in an eye-catching ascending gesture. In this way, the logo infuses representations with a consistently recognizable set of colors and forms. The partial or cropped logos engage the mind of the potential consumer to see a portion of a sign and actively recall and construct the whole. For a successful logo, a mere glimpse of its colors on page or screen will suffice to signify Blue Mountains, as consumers who are familiar with the brand only require

Using the Blue Mountains Signature

Permissible logo variations

There are only four permissible variations of the Logo and one of them must be used wherever possible. When size and/or technical constraints compromise its visual quality the BM Typemark is to be used as described below.

* Please note – all uses of the logo must be approved by Blue Mountains Tourism Limited. (see Brand Management section).

4 Colour – Primary Signature
In the majority of cases, the brand will be printed by the 4 Colour (CMYK) printing process, it is always presented on a white background.

4 Colour – CMYK Reversed
In the majority of cases, the brand will be printed by the 4 Colour (CMYK) printing process. Occasionally it is necessary to print this mark on a coloured background.

The 4 colour reverse brand can reverse from the Brand 'Mountains' (PMS 647 or its 4C breakdown) or Black.

Greyscale
This logo is used when a single colour black is being used and percentage 'tones' of black may be used to differentiate the areas of the logo. There is no solid black version of the brand icon, where a solid black version is required the Typemark should be used.

Greyscale – Reversed
This logo is similar to the Greyscale logo and is used when the corporate mark is used in applications where tones of only one colour can be used.

Figure 1. (Color online) Using the Blue Mountains Signature (permissible logo variations). Source: BMTL (2004, p. 12).

a reminder for recognition to occur. Logos are "markers of the edge between the aesthetic space of an image or text and the institutional space of a regime of value which frames and organizes aesthetic space" (Frow, cited in Lury, 2004, p. 13). The institutional space of BMTL and its network of brand partners are marked by the Blue Mountains brand signature, an aesthetic expression that depicts the physical space upon which this brand is then projected.

Color is a major element used in BBM. The Manual establishes a color palette which it characterizes as being directly derived from the natural landscape it symbolizes. An illustration shows how the varied features of the landscape were used to inform the brand colors (Figure 2). The accompanying text explains this palette:

> ...is drawn from the colors of the regional environment, and from the four seasons. The grid below [the palette] shows schematically what the grid at the left [the photographs] shows literally...the Blue Mountains glow with colors both vibrant and muted, complementary and contrasting. (BMTL, 2004, p. 19)

The selection of particular elements of the landscape to represent the regional environment and the seasons are necessarily limited in range, reproducing a stereotypical, traditional range of iconic birdlife, gum leaves, floral details and mountain views. Although described as muted and vibrant, the primary palette emphasizes boldness and saturated color in a way that owes more to the attention-seeking boldness of web pages than the hazy blue of the horizon. What is credited as indicating a year-round regional spectrum is necessarily a limited interpretation of color in the landscape. Local historian Jim Smith argues that the mountains "have many more than four" seasons, with migratory patterns and the flowering of specific plants marking numerous climatic and temporal cycles for those attentive enough to read them (Smith, 1988, p. 200), a level of subtlety and intimate knowledge of the landscape that remains unrepresented. Ultimately, the many colors and seasons of the landscape are selectively distilled as a highly regulated palette suited to the surfaces of twenty-first century marketing.

Font styles are determined by the Manual and are intended to communicate the brand personality. The Blue Mountains Brand has its own corporate font, Frutiger, which is described as being "contemporary [...] friendly and communicative [...] and] legible" (BMTL, 2004, p. 22). The authors claim that the font "gives the City a distinctive personality in its marketplace" (BMTL, 2004, p. 22). Other complementary fonts are specified for headings (Rotis Semi Sans) and for in-house publications such as reports and faxes (Arial). This level of specification indicates the control and continuity demanded by the branding process, which attempt to promote their idea of place identity to the letter, literally, by prescribing the visual appearance of text.

The brand provides rules for the content of text as well as its visual appearance. The Manual requires that "Copy and design must be integrated into the communication to deliver on the wider brand positioning" (BMTL, 2004, p. 11). This rule is followed in the headings used in BMTL marketing publications, which exude "upness" with phrases such as "welcome up" and "feels different up here" (BMTL, 2004, p. 36) (Figure 3).

Figure 2. (Color online) The color palette. Source: BMTL (2004, p. 18).

Figure 3. (Color online) Examples of brand graphic layout, typography and "elevated" copy. Source: BMTL (2004, p. 36).

The way in which the various elements are composed is as strictly scripted as the elements themselves. Layout guidelines specify how a sense of upness is to be conveyed by graphic composition (BMTL, 2004, p. 35). It is argued that typography and layout "play a powerful role in the creation of brand equity" (BMTL, 2004, p. 35). Stepped type and an "ascending" vignette (see the layout of words in Figure 3) are used to visually reinforce the upness message (BMTL, 2004, p. 35). The Manual states that these devices, when combined with visually standardized content, create an "authentic experience of the BM offering" (BMTL, 2004, p. 35).

Visualizing the Brand: The Photographic Strategy

> Photographers in the Blue Mountains and the distributors of their images have been powerful and selective mythmakers. (Snowden, 1988, p. 156)

Landscape photography has a rich history in the Blue Mountains, and BBM is the latest mythmaker to use photography to shape perceptions of this environment. The history of photography of the Blue Mountains is a significant topic in its own right, having been the subject of academic research and literature (Burke, 1988; Falconer, 1997; Snowden, 1988; Thomas, 2004). Whether critiquing the commercialization of the landscape through images, celebrating these images, or drawing artistic inspiration from them, all agree that photography has both reflected changing perceptions of the landscape and informed them. Since the 1860s, shifts here in landscape photography generally reflect the shifts in landscape perception as well as changes in photographic technologies that allowed new visions to be captured and distributed, consistent with Cosgrove's assertion that "the aesthetic conventions of landscape have been continuously reinforced by developments in mechanical and prosthetic vision" (2003, p. 257).

Early Blue Mountains photography was influenced by sublime mountain photography, which was in turn influenced by landscape painting traditions. It has been argued that the early experiences of photographers wishing to recreate such compositions were generally frustrated by the Blue Mountains terrain with its characteristic valleys and escarpments instead of towering peaks (Thomas, 2004, pp. 225–226). As infrastructure and technologies changed around the turn of the nineteenth century, so photographic images changed with them. Scenic lookouts on cliff tops were developed and promoted, and the panoramic elevated view was established as the pictorial norm, particularly for amateur photographers. During the mountains' busiest tourism decades prior to WWII, standard commercial imagery generally consisted of expansive panoramas from elevated viewpoints, capturing distinctive sandstone formations or waterfalls. Some images featured diminutive figures in the foreground, however increasingly the typical view itself was free of human presence, thus "creating and preserving and distributing an ideology of nature untouched" (Snowden, 1988, p. 137).

It is within this historical context that the Manual specifies photographic techniques aimed at reflecting the brand's own "unique Blue Mountains experience and perspective" (BMTL, 2004, p. 25). The Manual stipulates a range of commercial landscape imagery composed with the stated aim of changing the existing cultural image of the place. Four photographic styles form part of the "visual richness" of the brand strategy: Dynamic perspective, Intelligent Experiences, Macro Details and Standard Panoramic. These will now be described and analyzed in detail.

The primary photographic style stipulated in the Manual, "Dynamic perspective," describes how images will "[a]lways show people engaging with the natural environment in ways that challenge the traditional understanding of the [Blue Mountains] offering" (BMTL, 2004, p. 27). Images of people actively engaging with the environment should be shown in extreme perspective, characterized by acute angles, subjects viewed from above, below or contrapuntally (Figure 4). The strategy literally encourages the viewer to perceive the mountains from an unconventional point of view; for example, the Manual features a photograph of a man with a boy perched on his shoulders, peering up from a forest of tree ferns – a combination of elevated viewpoint and elevated content.

This compositional approach has been inspired by the vertiginous experiences the landscape affords, and "by the paintings of the celebrated Australian artist, William Robinson" (BMTL, 2004, p. 27) (Figure 4). The appropriation of a recognized Australian landscape painter's perspectival approach is significant. Robinson's distinctive "multi viewpoint" style (Robinson, cited in Klepac, 2001, p. 105) has been described as one where "the viewer is drawn into the natural world as an active participant" (National Gallery of Australia [NGA], 2006). Unlike panoramic detached views, his work evokes the sensory qualities of immersive landscape experience. Robinson states, "I am trying to achieve the non-static, the total relationships of moving elements, as we would sense, as we do, when we are in the landscape itself" (cited in Klepac, 2001, p. 105).

Such descriptions align with the experiential and sensory intent of the brand; several defining aspects of Robinson's landscape painting are, however, more difficult to reconcile with the brand and the landscape image BMTL constructs. First, Robinson's landscape paintings are not of the Blue Mountains, but are based on his intimate knowledge of his surroundings in Queensland, many hundred kilometers north. Robinson has been described as a regional painter because his work "is absolutely an art of place," and he lives in the landscape he paints, a fact "central to his work" (Fern & Robinson, 1995, p. 17 and 59). The brand's self-defined aim to communicate what makes the Blue Mountains "unique" can be interpreted as inconsistent with an outside artist's interpretation of a remote landscape.

Second, the character of Robinson's vertiginous multi viewpoint evokes aesthetic responses antithetical to the upbeat tone of the brand. Fink makes the connection between Robinson's work and Edmund Burke's notion of the sublime – a view more associated with anguish than upness and which "is often more grotesque than beautiful" (2001, pp. 13–14). Such distinctions highlight the ambiguity of the language of landscape and of semiotic interpretation in general. The brand's vertiginous compositional style, *intended* to convey a sense of upness, could readily descend into signifying freefall: the relative certainty of the picturesque panorama is abandoned for the distortions of a less familiar and disorienting point of view.

Figure 4. (Color online) Examples of the "Dynamic Perspective" photographic strategy. The two non-photographic images (top left and center right) are reproductions of paintings by William Robinson. Source: BMTL (2004, p. 26).

Third, appropriating an original landscape expression in one medium and re-presenting it in another can also be problematic. In this case, the scale of Robinson's large landscape painting is at odds with the format of tour guidebooks and magazines. Size is important to Robinson's works (Fern & Robinson, 1995, p. 52), which rely on their grand scale to envelop the viewer and heighten the sense of being within the landscape. The scale of photography reproduced for BMTL is necessarily much reduced, as is its impact.

Finally, and perhaps most importantly, Robinson's compositional style cannot be dissociated from the content of his landscapes, which has little to do with the expressed intention of the brand strategy. Robinson:

> is that most deeply unfashionable thing – he is a religious artist. This is not to say that his art can be reduced to an expression of faith alone; but to ignore the faith underpinning the work is to miss its animating principle. (Fink, 2001, p. 13)

Robinson's work is a religious meditation and celebration. His work may be stylistically contemporary; however, the ideas expressed through that choice of style are traditional. By appropriating a personal and deeply religious landscape painting style for mass distribution in the global marketplace, the brand strategy negates and evacuates the original meaning in an individual artist's work and secularizes it. The branded landscape constructs a form of commerce-friendly nature-based spirituality to appeal to a wide market. As BMTL manager Kerry Fryer explains: "one of the things that came very strongly [in the market research] was the emotional aspect or the spiritual aspect that had nothing to do with religion – going to nature" (Fryer, personal communication, 2005).

The brand presents spirituality as a generic "oneness with nature" that is neither Christian, nor, notably, is it Aboriginal. The absence of Aboriginal spirituality in the brand further reveals its limited evocation of the "spiritual aspect" of landscape that Fryer makes reference to, and indicates the wider problem of representing an imagined "unified community." Any sense of the spiritual qualities of this landscape as understood by its original inhabitants has been treated with ambiguity historically (see Smith, 1991; Thomas, 2004); the brand contributes to a tradition of the marginalization of Aboriginal landscape meaning and experience by neglecting such associations.

This is arguably to the detriment of the brand's own aims, since Aboriginal stories associated with the Blue Mountains are ones which encourage "an engagement with country" (Thomas, 2004, p. 92) on a deep experiential and localized level. Nevertheless, complex Aboriginal narratives do not lend themselves to the tagline brevity of brand communications, as they fail to fit into a three-word phrase. Nor do they find stylistic expression in a visual compositional convention or motif. In short, the BBM photographic strategy of "Dynamic Perspective" illustrates the brand's tendency to selectively misappropriate some landscape narratives and overlook others. It also reflects a conservative bias, since despite its claims to "show landscapes in non-traditional ways" the brand nevertheless conforms to the well-established tradition of portraying nature as an aesthetic object to be consumed. The claim that the BBM landscape is "non-traditional" is accurate in the unintended sense that the perspective of the traditional owners is overlooked.

The second photographic style for the brand, called "Intelligent Experiences" (BMTL, 2004, pp. 28–29), demonstrates how the brand is intended to change people's engagement with the landscape by emphasizing relationships between people and the natural environment (Figure 5). Intelligent Experiences photography

> seeks to visually demonstrate an equal weighting of human beings with the natural environment. People take the foreground and dynamically engage with nature in uplifting, positive ways; touching, breathing, embracing, eating, relaxing, enjoying. Just as Dynamic Perspective is inspired by the overwhelming experience of "looking out / up / down," Intelligent Experiences balance the impact of the natural world on individuals, couples, families, culture, cuisine, hospitality, adventure, spirituality and so on. (BMTL, 2004, p. 29)

The strategy of representing the wilderness as populated is one that signifies a departure from the norm. The brand delivers a "motivating message" intended to entice people into the landscape by making their potential presence explicit (BMTL, 2004, p. 7). In the discourse of place branding, an uninhabited place is a failed product; thus the need to produce imagery of the landscape product being successfully and happily inhabited and therefore consumed. An uninhabited wilderness may presuppose our presence, but an inhabited, branded wilderness demonstrates exactly what experiences are on offer. The photographic strategy thus:

> Gives equal weighting to environment and people to *demonstrate* engagement with the unique atmosphere of the BM [Blue Mountains]. It is in this range of images that cultural, social, recreational, spiritual or commercial aspects of the BM offering may be drawn out [...The strategy] Shows the landscape / Views / "Mountain Icons" sharing the frame equally with people clearly enjoying themselves and the environment in which they are situated. (BMTL, 2004, p. 29, emphasis added)

The Manual predetermines for brand partners both the content of landscape photography and its composition within the picture frame. The semiotic connection between signified and signifier is spelled out. Here, the "equal weighting" of human activity and nature is signified through the two subjects "sharing the frame equally." Similarly, the dynamic perspective and intelligent experiences approaches emphasize movement through a combination of content (active figures) and form (dynamic composition). Examples include photos of a young woman riding a mountain bike near the edge of a vertical cliff, others feature a group of young couples bush walking.

Although this strategy introduces people into the photographic landscape, the human element of the images is limited to the transitory physical presence of a small number of individuals, itself a limited view of what an inhabited landscape entails. Most visitors to the Blue Mountains National Park follow the extensive network of walking tracks that cover the landscape. Those shown in BMTL brochures and web pages, however, are depicted in untouched wilderness without a walking track in sight. The fact that only a few people are seen at any one time in these images reinforces the idea of the landscape as a wilderness without disturbance, despite the fact that the National Park walking tracks have historically been traversed by large

Intelligent Experiences

This is the second most important photographic style. This photographic approach seeks to visually demonstrate an equal weighting of human beings with the natural environment. People take the foreground and dynamically engage with nature in uplifting, positive ways; touching, breathing, embracing, eating, relaxing, enjoying.

Just as Dynamic Perspective is inspired by the overwhelming experience of 'looking out/up/down', Intelligent Experiences balance the impact of the natural world on individuals, couples, families, culture, cuisine, hospitality, adventure, spirituality and so on.

The desired outcome for each seasonal shoot will be a set of shots using the 'Intelligent Experiences' style that:

- Gives equal weighting to environment and people to demonstrate engagement with the unique atmosphere of the BM. It is in this range of images that cultural, social, recreational, spiritual or commercial aspects of the BM offering may be drawn out
- Brings the culture of the Mountains into the picture by including images of architecture, festivals, cafe society, hospitality etc
- Shows the landscape/Views/Mountain Icons' sharing the frame equally with people clearly enjoying themselves and the environment in which they are situated
- May have an 'UP' orientation, but should not compete with those images that are specifically Dynamic Perspective
- Employ crisp focus, saturated, 'in your face' images.

Figure 5. (Color online) "Intelligent Experiences" photographic strategy. Source: BMTL (2004, p. 29).

groups of people, and that walks along some of the easy tracks near population centers and major lookouts are used by tens of thousands of walkers every year.

The brand's third photographic strategy, "Macro Details," stipulates that communications should use close-up photographic views of landscape color and texture (Figure 6). Macro photographs of vegetation are intended to evoke how "the senses are engaged by the closeness of the natural world" while acting as a "Showcase for the flora / textures of the region." By zooming in on elements of the environment, it is suggested that the brand will provide "a fresh look at detail [...] presenting the landscape in all its uniqueness and beauty" while "giv[ing] marketing material extra depth" (BMTL, 2004, p. 33). This approach lends itself to current photographic and image reproduction technologies capable of powerful magnification and high resolution. Further, it is stated that Macro Details, and indeed all Blue Mountains brand photography, should employ crisp focus and saturated imagery (BMTL, 2004, pp. 25–33), thus moving away from hazy romantic notions of a Picturesque misty mountain landscape. Like the logo, they should be vibrant and bold. This representational strategy marks a significant move away from traditional depictions of an imposing landscape, a mysterious-yet-familiar panorama considered too distancing and remote for today's marketplace.

The brand's fourth photographic strategy, "Standard Panoramic," employs traditional panoramic views with a contemporary twist. The Manual recommends that if panoramic landscape imagery is used, it feature "NEW views of BM at different times of the day" (BMTL, 2004, p. 31, emphasis in the original). The two example panoramas in the Manual are atypical of picturesque panoramas in some ways, as a rainbow appears in one image, the horizon is absent in another, and neither image is framed by vegetation. These images nevertheless retain some traditional characteristics of panoramic composition, with people shown in the foreground and off to one side of the frame, their scale diminished in relation to the vast landscape beyond.

Figure 6. (Color online) Example of "Macro Detail" photographic strategy. Source: BMTL (2004, p. 2).

Overall, the brand's photographic strategy intends to signify elevation on "environmental, cultural, spiritual and commercial levels" (BMTL, 2004, p. 25). Despite the potential depth of these multiple levels of signification, the ability of commercial photographic composition to represent a full range of environmental, cultural and spiritual perspectives is limited. The strategy emphasizes variety – viewpoints, times of day, scales, activities in the landscape – while at the same time strictly determining what that variety must and must not include. Some stylistic innovations introduce a level of intimacy with the natural landscape, partially reversing the twentieth-century dominance of the panorama. However, by limiting the compositional variety and content of landscape images, the brand institutionalizes a new "mono-dimensional view" rather than expanding upon the existing one.

Applying the Vision: Controlling the Image

The Manual is a management tool intended to guide the visual communication of the brand by numerous partners. Brand manager BMTL is intent on having as many expressions of the brand as possible. Businesses are actively recruited to join the branding effort, for example a membership campaign begun in late 2006 ran the line "Get behind the brand and the brand will get behind you" (BMBA, 2012). Once businesses and other partner organizations are recruited, they are required to follow the rules of the Manual to the letter. Like a franchiser, brand owner BMTL has the legal capacity to enforce copyright, controlling any unauthorized or non-conforming application of the brand: "All copyright and intellectual property associated with the brand and the elements of the brand belong to BMTL" (BMTL, 2004, p. 38). A section of the Manual dedicated to stipulating the all-important copyright and managerial conditions states: "Where brand elements are used by brand partners, these must be approved by BMTL. Artwork must be submitted for approval, and where design elements such as the sub-graphic are used, a separate application and approval from BMTL is required" (BMTL, 204, p. 38).

BMTL also detail how they will implement brand control within their tourism plan. Objective 1.1.7 requires BMTL to "[r]egularly audit the content of the key image influencers of third parties and attempt to modify this where it is out of alignment with the optimal branding defined in the brand style guide [Manual]" (GTL, 2004, p. 9).

The Manual leaves little room for error or interpretation, as its highly prescriptive instructions spell out how the logos, fonts and photographs are to be reproduced uniformly by all. For example, the standard application of the color palette is controlled by accompanying instructions for color reproduction, whether in print, on-screen, embroidery and even wall paint (BMTL, 2004, pp. 19–20). Examples of correct and incorrect applications are illustrated, with the text insisting that variations must "NEVER" (BMTL, 2004, p. 12, emphasis in the original) be made (Figure 1). This control is justified by BMTL on the grounds that incorrectly applied visual communications "will weaken or damage the integrity and impact of the identity" of the brand (BMTL, 2004, p. 15).

Brand management seeks a consistent and mutually reinforcing image of the Blue Mountains landscape distributed by multiple organizations, thus raising the prospect of a uniform landscape of ideas. Whereas in the past local business operators were free to describe the landscape as they saw fit, they are now being requested to reiterate predetermined, dominant themes. It should be acknowledged that it is unlikely that one brand could enforce a "single-minded" visual image across an entire community (unlike a single organization or company brand). A systematic audit of all landscape imagery being produced in the Blue Mountains since 2004 is beyond the scope of this study. However, amid the many examples of conforming logos and layouts, some brand partners appear to have broken the pictorial rules, for example by persisting in using photography featuring traditional panoramas and iconic views. Furthermore, others outside the brand continue to independently produce landscape ideas and images; being a brand partner is voluntary, so many businesses have their own communications strategies, and local artists and tourists continue to depict the environment in their own ways – from blogged photographs through to exhibitions and fiction – without a brand to prescribe how they do so. Nevertheless, BBM's *aim* to create a consistent and repetitive place image is in itself significant, and is brought closer to realization through a legally enforceable (copyright) system of brand management.

Conclusion

This analysis has shown that BBM communicative strategies seek to systematically control the image of the Blue Mountains landscape. The brand Manual is specific about the "desired outcome(s)" of the strategy in terms of all image composition and content (BMTL, 2004, p. 27). In this system, the natural environment is never a neutral pictorial element, but is employed instrumentally as a means of promoting a "single-minded, compelling and unique" version of the place to outsiders and to locals alike.

The brand selectively interprets existing characteristics of the landscape as well as importing others, with the resulting combination conforming to the perceived desires of targeted market demographics. The strategy makes use of the symbolic potential of the landscape to translate abstract concepts – "values," "personality," and "sustainability" – into tangible place images. It comprehensively specifies how the landscape is represented in terms of color, texture, form, temporality, human interaction and experience, personality and narrative conventions, spelling out what messages the various image styles or content are intended to convey. It marks a conscious departure from the traditional depictions of an imposing landscape, rejecting distant panoramas and over-familiar iconic sites in favor of more immediate, people-focused narratives. Together the logo, color scheme and highly prescriptive photography combine to communicate the upbeat, landscape-led place identity.

The imagery produced by the brand reflects characteristics of the Blue Mountains' natural environment, but it equally reflects the demands of the existing media environment. Evocations of the landscape vary depending on the specific qualities of the medium used, with the specific qualities of various media being exploited for their own greatest effect. Logos demand boldness, copywriting demands brevity, photographic composition exploits zoom lenses and high-resolution reproduction quality, and color is standardized to suit current printing and digital technologies. Overall the Blue Mountains landscape is rendered colorful and condensed.

Central to BBM is the notion of an elevated landscape, which is used to metaphorically – and then translated visually – to reflect abstract concepts such as business prosperity, community spirit and emotional states as well as promoting the natural landscape setting. This implies that socio-cultural constructs (i.e. community values and economic competitiveness) constitute a natural extension of the elevated landscape, reifying associations between the landscape and notions such as nature, purity, emotional elevation, financial / business elevation, and community unity. Insisting that a place construct is seemingly "natural" lends an air of inevitability to the work, concealing the selective process involved.

Although BBM challenges certain pictorial stereotypes of an uninhabited scenic mountain landscape, the portrayal of the environment as naturally pure and uplifting perpetuates existing conventional notions of the mountains as a retreat from contemporary life. The BBM strategy simplifies the image of a region which is characterized by a rich pattern of human settlements surrounded by National Park areas. While the environmental benefits and challenges that arise from this park proximity continue to be enjoyed and negotiated by residents and local government, it is difficult to gain a sense of this complexity from the landscape representations prescribed by BMM.

The emphasis given to the "naturalness" of the landscape by BBM implicitly denies and negates important cultural landscape associations such as the history of Aboriginal occupation and the recent efforts to reconcile and acknowledge traditional ownership of land. The absence of Aboriginal narratives, along with the appropriation of William Robinson's visual style – without his artistic content – serve as examples of how the brand's visual communications are bright and bold on the surface but lacking depth. It seems that as long as brands insist on being "single-minded" there will be little room for anything other than a natural and secular Blue Mountains, as complex narratives and tensions are incompatible with branding aims and structures.

In summary, the framework that the brand sets for the representation of the landscape is an exercise in calculated aesthetics, whereby the form and content of landscape images of various kinds is measured to achieve the greatest differentiation and impact which technologies allow. The result of this calculated aesthetic system, with its taglines and magazine-format brevity, is a reduction in the complexity of landscape representations and a perpetuation of nature stereotypes. The declaration of an all-encompassing "Vision" for the region, the all-purpose tagline *Elevate Your Senses*, and the drafting of principles relating to the region's identity, values and personality are all intended to guide perceptions of the region toward a single coherent, repetitive – and relentlessly positive – theme. When "new" landscape concepts and compositional forms are introduced which seek to promote a more sensorily engaged experience of nature, these do not challenge the underlying value system that presents the landscape as an object for consumption.

This description of BBM, and of place branding as a practice in general, foregrounds its highly ideological character, as ideologies "offer ordered, simplified versions of the world; they substitute a single certainty for a multiplicity of ambiguities; they tender to individuals both an ordered view of the world and of their own place within its natural and social systems" (Baker, 1992, p. 4)

Attempting to order and control the expression of identity is place branding's raison d'être. The degree of planning and control that place brand strategists invest in the visual communication of brand identity is significant. This attempt at control and indeed legal ownership of expressions of landscape challenges the very notion of landscape as a subjective, and hence individually "owned," concept: In 1836 Ralph Waldo Emerson (Emerson, cited in Mitchell, 1994, p. 15) remarked that "landscape has no owner," thus distinguishing the exchange value of land-as-property from the poetic / symbolic property of landscape as a cultural expression. However, in the twenty-first century, the exercise of intellectual property rights by place brand managers makes this distinction less clear. Brand managers cannot copyright the idea of a landscape, but can copyright the material expression of that idea in the form of words, logos and photographs.

Alternatively, for the full potential of human-nature relations to be visually communicated, place branding strategies will have to be more flexible in their interpretation of "identity," and not quite so single-minded. After all, nature and place cannot be reduced to a singular essence. It is not the role of a single authority to prescribe the meaning of place for all, regardless of its content or expression.

The colorful logos and carefully composed photographs of place branding must form part of a much wider visual spectrum, one capable of constructing a truly unique version of nature and the possibilities it holds. By definition it is impossible to have an unmediated landscape or a pure "nature", as these are cultural constructs to begin with. In this sense the landscape is not more or less mediated, not more or less authentic, only more or less *consistently* constructed and enforced. Ultimately, it is with a diverse variety of depictions of natural environments – not standardized visual communications strategies – that "compelling" and "unique" places will emerge.

References

Anholt, S. (2004). Nation-brands and the value of provenance. In N. Morgan, A. Pritchard, & R. Pride (Eds.), *Destination branding: Creating the unique destination proposition* (2nd ed.). (pp. 26–39). Oxford: Elsevier Butterworth-Heinemann.

Arvidsson, A. (2006). *Brands: Meaning and value in media culture*. London: Routledge.

Baker, A. R. (1992). Introduction: On ideology and landscape. In A. R. Baker & G. Biger (Eds.), *Ideology and landscape in historical perspective* (pp. 1–14). Cambridge: Cambridge University Press.

Blue Mountains Business Advantage (BMBA) (2012). BMBA official web site. Retrieved May, 2012, from http://www.bluemountainsadvantage.com.au

Blue Mountains City Council (BMCC). (2011). *BMBA official web site.* Retrieved January, 2011, from http://www.bmcc.nsw.gov.au/

Blue Mountains Tourism Limited (BMTL). (2004). *Brand Blue Mountains manual* (Version 2). Katoomba: BMTL.

Burke, A. (1988). Awesome cliffs, fairy dells and lovers silhouetted in the sunset: A recreational history of the Blue Mountains, 1870–1939. In P. Stanbury (Ed.), *The Blue Mountains: Grand adventure for all* (pp. 99–117). Sydney: Leura, Second Back Row Press, The University of Sydney.

Cosgrove, D. (2003). Landscape and the European sense of sight—eyeing nature. In K. Anderson, M. Domosh, S. Pile, & N. Thrift (Eds.), *Handbook of cultural geography* (pp. 249–268). London: Sage.

Falconer, D. (1997). *The service of clouds.* Sydney: Picador.

Fern, L., & Robinson, W. (1995). *William Robinson: Roseville east.* NSW: Craftsman House in association with G + B Arts International.

Fink, H. (2001). Light years: William Robinson and the creation story. *Artlink, 21*(4), 12–17.

Global Tourism & Leisure (GTL). (2004). *Blue Mountains regional three year tourism plan and implementation program, 2004–2007.* Katoomba: BMTL.

Gold, J. R., & Revill, G. (2004). *Representing the environment.* New York: Routledge.

Hartig, K. (1987). *Images of the Blue Mountains: Research monograph series.* Sydney: University of Sydney Department of Geography.

Kavaratzis, M. (2004). From city marketing to city branding: Towards a theoretical framework for developing city brands. *Place Branding, 1*(1), 58–73. doi:10.1057/palgrave.pb.5990005

Kavaratzis, M. (2005). Place branding: A review of trends and conceptual models. *The Marketing Review, 5*(4), 329–342. doi:10.1362/146934705775186854

Klepac, L. (2001). *William Robinson: Paintings 1987–2000.* Roseville, NSW: Beagle Press.

Lindstrom, M. (2005). *Brand sense: Build powerful brands through touch, taste, smell, sight, and sound.* New York: Free Press.

Lury, C. (2004). *Brands: The logos of the global economy.* New York: Routledge.

Middleton, V. (2001). *Marketing in travel and tourism* (3rd ed.). Oxford: Butterworth-Heinemann.

Mitchell, W. J. T. (1994). Imperial landscape. In W. J. T. Mitchell (Ed.), *Landscape and power* (pp. 5–34). Chicago: University of Chicago Press.

Mommaas, H. (2002). City branding: The necessity of socio-cultural goals. In V. Patteeuw, T. Hauben, & M. Vermeulen (Eds.), *City branding: Image building and building images* (pp. 34–47). Rotterdam: NAi.

Morgan, N., & Pritchard, A. (2004). Meeting the destination branding challenge. In N. Morgan, A. Pritchard, & R. Pride (Eds.), *Destination branding: Creating the unique destination proposition* (2nd ed.). (pp. 59–78). Oxford: Elsevier Butterworth-Heinemann.

National Gallery of Australia (NGA). (2006). *Masterpieces for the Nation Fund New Acquisition.* Retrieved March, 2013, from http://nga.gov.au/AboutUs/press/masterpieces.cfm

Porter, N., & Bull, C. (in press). Conceptualizing, representing and designing nature: Cultural constructions of the Blue Mountains', Australia. In E. Carr, Eyring, S. & Wilson R. G. (Eds.), *Public nature: Scenery, history and park design.* Charlottesville, VA: University of Virginia Press.

Scharma, S. (1995). *Landscape and memory.* London: Harper Collins.

Schmitt, B. (1999). *Experiential marketing: How to get customers to sense, feel, think, act, and relate to your company and brands.* New York: Free Press.

Smith, J. (1998). *Blue Mountains national park walking track heritage study.* Sydney, NSW: National Parks and Wildlife Service.

Smith, J. (1988). Blue Mountains myths and realities. In P. Stanbury (Ed.), *The Blue Mountains: Grand adventure for all* (pp. 185–202). Leura, Sydney: Second Back Row Press; The University of Sydney.

Smith, J. (1991). *Aboriginal legends of the Blue Mountains.* Wentworth Falls, NSW: Jim Smith.

Snowden, C. (1988). The take-away image: Photographing the Blue Mountains in the nineteenth century. In P Stanbury (Ed.), *The Blue Mountains—Grand adventure for all* (pp. 133–156). Leura, Sydney: Second Back Row Press; The University of Sydney.

Thomas, M. (2004). *The artificial horizon: Imagining the Blue Mountains.* Carlton, Vic.: Melbourne University Press.

The Nature of *Time*: How the Covers of the World's Most Widely Read Weekly News Magazine Visualize Environmental Affairs

Mark S. Meisner & Bruno Takahashi

Scholars of environmental communication acknowledge the importance of visual representations in shaping perceptions and actions in relation to environmental affairs. Unlike with other media, including newspapers, television and film, research on the visualization of nature and environmental issues in magazines is rare. This study focuses on the covers of Time *magazine, one of the world's most influential news weeklies. A dataset that includes all relevant covers from 1923 to 2011 is examined using a combination of quantitative and qualitative content analysis to analyze the visual representation of nature and environmental issues. The results show that the presence of environmental issues and nature on the covers has increased over the decades. Furthermore,* Time *takes an advocacy position on some environmental issues, but it is a shallow one that is weakly argued through less-than-engaging imagery and fails to offer much in the way of solutions or agency to the reader.*

Nature and environmental themes are intrinsically visual, and human perception and understanding of environmental affairs are deeply influenced by the visualizations created for the media. Images can also move people to action, as is well known in the environmental movement (Dale, 1996; DeLuca, 1999). Thus, it is important to

---

Mark S. Meisner is the Executive Director of the International Environmental Communication Association (IECA) and a visiting scholar at the Center for Environmental Policy and Administration at the Maxwell School at Syracuse University. Bruno Takahashi is an Assistant Professor in the School of Journalism and Department of Communication at Michigan State University.

understand the imagery of environmental affairs in order to understand its role in shaping ideologies of nature and, ultimately, what is done to address the social problems underlying environmental issues.

Interest in studying the representations of nature and environmental issues in the mass media has increased in recent decades. Key examples of recent work in this area include those by Lester (2010), Hansen (2010), Boykoff (2011), and Doyle (2011). These, in turn, built on earlier works by Wilson (1991), Hansen (1993), Hannigan (1995), Dale (1996), Anderson (1997), DeLuca (1999), Shanahan and McComas (1999), Mitman (1999), Allan, Adam, and Carter (2000), Bousé (2000), Ingram (2000), Brereton (2004), and Corbett (2006), among many others. These efforts provide a perspective on which issues are more salient and how they are framed, as well as on the myriad of factors that shape news coverage and media representations. Furthermore, many of these have included consideration of visual representations.

Compared to other media such as newspapers, television, and film, magazines have had relatively little attention in this environmental communication literature and even less from a visual communication standpoint. This study considers magazine covers as a prominent vehicle to visually communicate about nature and environmental affairs to the public. More specifically, we focus on covers of *Time*, considered one of the world's most well-known magazines.

As McQuail (2005, p. 31) has noted, magazines have generally been neglected in communication research, likely because of their diffuseness and uncertain impact. Among magazines, though, very few have the iconic global status and recognition of *Time*, historically the most widely read and influential weekly news magazine (Angeletti & Oliva, 2010). With the media landscape rapidly changing with the transition to a continuous news cycle and digital distribution, it is possible to question the ongoing relevance of weekly news magazines. Nonetheless, the Pew Research Center's Project for Excellence in Journalism continues to regard *Time* as the strongest title in this category (Matsa, Sasseen, & Mitchell, 2012). With an approximate annual US paid circulation of 3.3 million, *Time* has more than double the *New York Times*' paid circulation (Matsa et al., 2012). According to *Time*'s own figures, "more than 40 million people around the world interact with *TIME* every week – a number that includes 25 million readers, 18 million unique monthly visitors on TIME.com, 2 million Twitter followers and 850,000 mobile users" (Stengel, 2010).

Time has also shown a long-standing interest in environmental affairs, having featured a number of special issues and prominent cover stories about environmental degradation generally and climate disruption specifically. Notable among these are the January 1989 "Planet of the Year" issue and the April 2000 and August 2002 issues, both titled "How to Save the Earth." Much of this work was led by *Time*'s then environmental correspondent Eugene Linden (2006), the author of *The Winds of Change: Climate, Weather, and the Destruction of Civilizations* (see http://www.eugenelinden.com/bio.html).

Our desire to study *Time*'s visual representations of environmental affairs was motivated by this knowledge and by our impression that *Time* is a somewhat more progressive publication that appears to favor a kind of journalistic advocacy with respect to environmental issues. Thus, to study how a high profile, visually prominent, and sympathetic news magazine visualizes environmental affairs is an opportunity to consider current best-case scenarios that can be expected from a mainstream publication.

Just to be clear, we note that several concepts comprise our definition of environmental affairs. These include environmental *issues* (e.g., global warming), which we see as symptoms, and *problems* (e.g., excessive fossil fuel burning), which are akin to underlying causes. We also include *players* in environmental debates (e.g., environmentalists) and *actions* to address issues and problems (e.g., tougher fuel efficiency requirements). Generally speaking, we follow a constructionist perspective on these matters (Best, 1995; Hannigan, 1995).

Furthermore, we are of the view that a key aspect of looking at representations within environmental discourse is considering the broader discourse of nature (Wilson, 1991). Environmental discourse concerns itself with the problematic relationship between humans and the rest of nature, yet this is a subset of the discourse of nature which concerns itself with what nature is, how it works, and what it means to humans. Therefore, we also include representations of *aspects of nature*, including natural processes, in the representation of environmental affairs.

What do we mean by "nature?" This unassumingly complex question has been taken up repeatedly within environmental thought. Recognizing the meaningful tension between a nature that includes humans and their creations and a nature that does not (Evernden, 1992; Lindahl, 2006; Soper, 1995), we have opted to use the word in the latter sense in this paper. So when we say "nature," we mean other-than-human-nature, or greater-than-human-nature. That said, we believe humans are, in fact, part of the natural world, even if much of what we do and think says otherwise. Furthermore, what we understand as "nature" or "natural" is culturally and historically contingent. In other words, the meaning of nature is socially constructed.

Covering Environments

Several threads of literature converge in this paper, including work on magazine covers, on *Time* magazine, and on the visualization of environmental affairs.

In a review of the state of research of magazine covers, Johnson (2002, p. 8) argued that there has been limited interest in studying such iconic popular culture artifacts. In advocating for the need to advance this area of study, she reasoned that "Magazine covers not only offer information about what's inside a particular issue, they also provide significant cultural cues about social, political, economic, and medical trends." Sumner (2002) tested this argument in an analysis of the degree to which the covers of *Life* magazine acted as a "cultural artifact" and/or a "marketing tool." The results suggested that *Life* magazine, in general, failed in performing either function. Nevertheless, even if magazine covers do not precisely reflect social, cultural, or political trends, their extended societal reach (especially in the USA), provides them with an important role in presenting and legitimizing issues. In this respect, Cerulo (1984) identified the importance of the usage of shared symbols in two American magazines, *Life* and *Look*, since these symbols can be easily internalized by individuals and affect understandings of issues and even behavior.

Studies of visual representations, including on magazine covers, have often focused on political and gender issues. Political scientists have analyzed the photographic portrayals of politicians (especially during campaign years) in newspapers and magazines (e.g., Moriarty & Popovich, 1991; Waldman & Devitt, 1998). Studies on representations of women on magazine covers usually report a shortage of covers representing women. Moreover, women are objectified, suggesting a stereotype that is incongruous with social norms (Leath & Lumpkin, 1992; Pompper, Lee, & Lerner, 2009).

Within the scholarship of magazines' pictorial representations, there has been a strong focus on *Time* magazine. Scott and Stout (2006, p. 2) argued the need to focus on *Time* because it reflects social trends, reveals public perceptions about historical events, and because of "...its dominance as a news vehicle in the arena of public discussion." In their analysis of religious depictions on the covers of *Time* during a decade-long period, the study found a focus on private over institutional religious practices and a synthesis with other themes in the arts, entertainment, and science. The use of certain images over others can have a profound effect on social and political realities, as suggested by Perlmutter (2007, p. xxvi), who argued that *Time* magazine used images "...to skew, recode, redirect, or overturn for persuasive ends" the public image of China.

Other studies of *Time*'s covers reveal a variety of topic areas. For instance, Christ and Johnson (1985) reviewed all the *Man of the Year* covers from 1927 through 1984 in terms of gender, age, nationality, citizenship, and occupation. A follow-up article found that women are portrayed infrequently on the covers of the magazine (Johnson & Christ, 1988). Similarly, Bates (2011) found a significant decline in the presence of "public intellectuals" on the covers since the 1960s.

Finally, in one of the few articles focusing on environmental issues, Loi (2010) analyzed *Time*'s coverage of climate disruption between 2004 and 2007. The study revealed that the coverage increased during that period, that there was a shift from uncertainty to certainty, and that climate disruption was "mainstreamed" as it was discussed in the context of other issues. However, Loi did not focus on the visual portrayal of the issue.

With respect to research on environmental communication, Hansen (2011, p. 17) has said, "...rarely has the analysis of visuals or the visualization of the environment been the main focus of analysis, or indeed the subject of systematic or longitudinal analysis." Nevertheless, Hansen (2011) recognized promising signs of a shift, with more studies analyzing the visuals of newspapers (DiFrancesco & Young, 2011; Seppänen & Väliverronen, 2003; Smith & Joffe, 2009), campaign materials (Doyle, 2007; Manzo, 2010a), image banks (Hansen & Machin, 2008), and television news (Lester & Cottle, 2009). To these, we would add photography (Peeples, 2011), film (Ivakhiv, 2010), online news sites (Slawter, 2008), and magazines (Remillard, 2011; Todd, 2010).

Early studies of visual representations of nature and environmental issues in magazines looked back to the first days of the modern environmental movement in the USA, the 1960s and 1970s. The movement was reflected in the number of new magazines and older magazines adjusting their content to focus on the environment between 1966 and 1975 (Schoenfeld, 1983). Howenstine (1987) also studied the coverage of environmental issues in the *New York Times*, the *Washington Post*, *Time*, and *Newsweek*, comparing the coverage in 1970 and 1982. Although the study found an increase in environmental coverage, there was also an increase in the use of economic/development stories.

The often dichotomous relationship between nature/environment conservation and development was examined by Remillard (2011), who analyzed a photographic essay by *National Geographic* magazine on the Canadian oil sands. The study revealed a tension between nature as a resource and nature as sublime. These representations of risk were framed within a Western perspective, clearly neglecting indigenous perspectives. Similarly, Todd (2010) argued in an analysis of *National Geographic's* presentation of Africa that an anthropocentric point of view predominated.

There have also been some studies analyzing the visual depiction of nature. For example, magazines' images of insects show that most magazines in various subject areas (i.e. science, home, women, and men) portrayed insects mostly in negative terms. Only nature magazines tended to present them more positively (Moore, Bowers, & Granovsky, 1982). Similarly, Kalof and Fitzgerald (2003) studied visual representations of dead animals in hunting magazines. They found a profound objectification of animals, far removed from poetic and taken-for-granted notions of love and affection toward nature and animals. In trying to explain the functions performed by photographs of biodiversity, Seppänen and Väliverronen (2003, pp. 80–81) suggested four roles within the context of the UK's *The Times* newspaper: "(1) concretize a complex scientific problem; (2) construct social relationships between different actors; (3) provide an opportunity for affective involvement; and (4) produce a 'reality effect.'"

The recent interest in analyzing visual representations of environmental issues has been driven mostly by the rise of climate disruption as a social issue. DiFrancesco and Young (2011) and Smith and Joffe (2009) found that newspapers in Canada and the UK, respectively, have mostly presented climate disruption visually using a "human face." In Canada, the visual representations of impacts are less prominent than in the UK. Smith and Joffe (2009) argued that the British press has used visual imagery in three ways: to bring the threat closer to home, to personify climate disruption, and to graphically represent it. They argued that such imagery could influence the salient information in people's memory and their emotional engagement, as well as their engagement in climate disruption risk actions. This focus on the visual representations of climate disruption has also expanded to television, where news stations in six Western countries and on satellite tended to use spectacular and visually resonant images (Lester & Cottle, 2009). Nicholson-Cole's (2005) study of people's visual conceptions and feelings about climate disruption revealed that such conceptions were associated with visual representations of the issue in the media.

These visual studies have mostly presented critical views of the ways the media and advocacy groups have used visuals to communicate the importance of the issue. Similarly, Manzo (2010b) discussed the need to move beyond fear appeal messages and polar bears symbols and, instead, provide more inspirational visuals. Moreover, Manzo (2010a) discussed the importance of embedded geopolitical issues in the representations of vulnerability in climate disruption communication campaigns. Along the same lines, Doyle (2007) discussed the challenges of visually presenting a long-term problem based on a temporal dimension (future effects) and its nonvisible nature. Finally, Hansen and Machin (2008), in their study of Getty images, also criticized the decontextualization of such images to suit marketing needs.

From this brief review of some of the most pertinent literature, we conclude that media visualizations frequently present conflicting images of nature, as well as somewhat weak and ineffective representations of environmental issues. There are also disagreements among these and other scholars mentioned earlier about the appropriate role for such images, and the basis by which they should be judged.

Questions and Methods

Our main research question is *How are environmental affairs represented on the covers of* Time *Magazine?* Four subquestions provide more precision and a framework for analysis:

(1) Which environmental issues, players, and actions are represented?
(2) How are the issues, players, and actions represented?
(3) Which aspects of nature are represented?
(4) How are the aspects of nature represented?

These questions were used to guide both a content analysis and a critical qualitative analysis of the covers, as described below.

Data Collection

Time magazine maintains an archive of all of their American edition covers from 1923 to the present at www.time.com. As of the 26 December 2011 issue of the magazine, there were 4653 covers in the archive. The archive is searchable by keywords which *Time* staff have used as metadata for the covers. To find the covers that were relevant to our study, we scanned the keywords and searched for those related to environmental affairs. These included (no. of hits in brackets) agriculture (20), animals (26), disaster (24), earth (7), energy (35), environment (66), food (30), global warming (15), nuclear (43), oil (25), pollution (4), population (2), water (5), weather (16), and wildlife (10). Then we visually scanned the whole archive for additional relevant covers, ignoring small, incidental representations of nature that had nothing to do with the subject matter.

After removing duplicates, we had 312 unique covers. Of those, many that we found, using the keyword searches simply did not contain significant visual or textual representations of nature or environmental issues, often because they were framed differently. For example, of the 35 covers tagged by *Time* with "energy," only nine had any sort of environmental theme or frame. The rest were about blackouts, the cost of oil, oil companies, etc. We excluded those from our set. We also excluded covers with representations of aspects of nature that were either inconsequential to the overall image, fictional, extraterrestrial, or metaphorically being used to signify something else altogether unrelated to environmental affairs. Examples of such metaphorical uses include Wall Street bulls and bears, Republican elephants, Democratic donkeys, and male-chauvinist pigs. We are aware that such representations can play a part in the social construction of nature or aspects of it (Baker, 1993; Lawrence, 1993; Meisner, 1997 2005), but decided that including them would not help us answer our research questions. After applying these criteria, we were left with a total of 128 covers in our dataset (See Appendix for the complete list of issue numbers). This number represents just 2.75% of *Time*'s US covers.

Data Analysis

As indicated by the research questions, our interest was in understanding not only what was being portrayed on the covers but also on how it was being portrayed. We wanted to know how issues were being symbolized and how aspects of nature were being represented. So the analysis had to be part quantitative visual content analysis (Bell, 2001; Hansen et al., 1998), coding for the presence of environmental issues and aspects of nature, and part qualitative visual analysis of elements on the covers that contributed to the representations. This combined approach echoes that used by Leiss, Kline, and Jhally (1990) in their study of advertisements, a method that is "rigorous and systematic while also being sensitive to the multiple levels of meaning" (p. 225) in the texts.

We used QDA Miner software for the coding. In this process, we considered not just the visuals but also the text on the covers. In many cases, text provides information about how to interpret the visuals and vice versa. Because of the relatively small dataset, we both coded all 128 covers together, discussing and negotiating each disagreement, of which there were few.

We coded each cover for the presence of environmental issues (e.g., oil spill), how issues were represented (e.g., oil-soaked birds), aspects of nature (e.g., bird), representations of nature (e.g., nature as victim), types of people (e.g., scientist), and actions to address issues (e.g., improve technology). Covers could have multiple codes for each of these. All codes were open-ended, except those for representations of nature. For that, we adapted Meisner's (2005) categories and used "nature as victim/patient," "nature as problem," "nature as resource," "nature as research subject," "nature as human analog and/or companion," "nature as exemplar of fitness," and "nature as just itself," meaning no discernable characterization in relation to humans.

After the initial round of coding, we then grouped the open-ended data points and determined whether some codes and some categories could be combined.

Findings and Discussion

Given that the final dataset was smaller than anticipated (128 covers representing 2.75% of the archive, or a mean of 1.4 covers per year over the 89-year period), we are reluctant to make many claims regarding historical changes in representations. It had been our hope to be able to speak to changing issue focus and representations of nature over time, but there are not enough data points to make many such claims. Nevertheless, there are certainly some significant aspects to this data worth discussing.

The first thing to note about the representation of environmental affairs on the covers of *Time* magazine is just how few representations there are. Out of 4653 covers, only 67 (1.4%) are about environmental issues (see Table 1) and only 107 (2.3%) contain significant representations of nature (see Table 2). Perhaps this should not be surprising, considering how many different types of issues demand the media's attention. But given *Time*'s interest in environmental affairs, these numbers do seem low. Nevertheless, the total number of covers presenting either aspects of nature or environmental issues has followed an incremental trend throughout the decades (see Tables 1, 3, and Figure 1).

Issues Represented

Table 1 summarizes the frequencies of different types of environmental issues and when they appeared on the covers. In the 1960s and 1970s, there are a few key covers, including some on overpopulation, nuclear war, water shortages, severe air pollution,

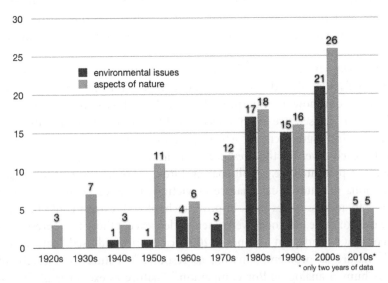

Figure 1. Numbers of covers with one or more environmental issue or aspect of nature.

Table 1 Types of environmental issues on *Time* covers by decade.

Issue	1920s	1930s	1940s	1950s	1960s	1970s	1980s	1990s	2000s	2010s[a]	Total
Covers with one or more environmental issue[b]											
Environment in general			1	1	4	3	17	15	21	5	67
Climate and weather						1	1	4	4	1	10
Coastal erosion							4	2	13	1	20
Drought							1		1		2
Flood							1				1
Forest fires								1	2		3
Global warming/climate disruption									1		1
Ozone depletion							1	1	9		11
Food and farming							1		1	1	2
Food safety							1		1	1	3
Food supply							1				2
Overpopulation										1	1
Pollution					1		1		1		3
Genetic pollution			1	1	2	5	13	7	2	4	35
Oil spill									1		1
Pollution general							1	1		2	4
Pollution – air					1			1		1	2
Pollution – water						1	2	1			5
Radioactive contamination				1		1	3				4
Radioactive contamination (war)			1		1	1	2	2	1		7
Toxic waste						1	2	2			8
Resource plundering							3				4
Deforestation					1		1	3	3	1	9
Energy shortage							1	1	1		3
Resource plundering in general								1	1	1	2
Water shortage					1			1	1		2
Wildlife and habitat								1			2
Habitat destruction						3	2	4	3	1	13
Species extinction						2	1	1	1		5
Wildlife loss								3	2	1	6
						1	1				2

[a]These numbers are for only two years of this decade.
[b]Some covers contain multiple issues, so simply totaling the other rows would give an inflated picture of the number of covers involved.

Table 2 Types of environmental issue representations on *Time* covers by decade.

Issue representations	1920s	1930s	1940s	1950s	1960s	1970s	1980s	1990s	2000s	2010s[a]	Total
Covers with one or more issue representations[b]			1	1	4	3	17	15	20	5	66
Humans					1	2	7	1	2	1	14
Suffering human							3	1	2	1	7
Skeleton person							2				2
Crowds of people					1	1	1				3
Traffic jam						1	1				2
Animals						3	7	4	4	2	20
Dead fish						1	2			1	4
Oil-soaked birds						1				1	2
Skeleton fish							2				2
Unharmed species at risk					1	1	3	4	4		12
Source of pollution				1	1	4	5	5	1		18
Burning oilfields								1			1
Effluent pipe						1					1
Mushroom cloud			1	1	1	1	2	2	1		9
Nuclear reactor						1	2	2			5
Smokestack fumes						1	1				2
Ecosystem effects					1		3		3	1	8
Dried-out field							1				1
Ocean wave									1		1
Oil on the water							1			1	2
Shrunken ice									2		2
Smog					1						1
Waves hitting cliffs							1				1
Other					1			1			2
Juxtaposition of nature and technology								1			1
Dripping tap					1						1
Metaphors							7	1	2		10
Anthropomorphic pollution monster							1				1
Burning hole								1			1
Burning skull							1				1

Table 2 (*Continued*)

Issue representations	1920s	1930s	1940s	1950s	1960s	1970s	1980s	1990s	2000s	2010s[a]	Total
Corn wrapped in money									1		1
Face of fear							1				1
Food in shape of question mark							1				1
Frying pan									1		1
Globe wrapped in plastic							1				1
Greenhouse							1				1
Heat ring							1				1
Graphi design						1		1	5		7
Black-and-white picture								1			1
Evocative colors						1	2	3	6		13
X-ray scene						1					1
Text					2	2	11	15	22	5	57
Literal text					1	1	9	12	19	5	47
Metaphorical text					1	1	2	3	3		10

[a]These numbers are for only two years of this decade.
[b]Some covers contain multiple issue representations, so simply totaling the other rows would give an inflated picture of the number of covers involved.

and perhaps the most interesting of all the covers, one featuring ecologist Barry Commoner (1970-02-02). However, it is not until the 1980s when noticeable numbers of environmental issues are featured. Pollution issues dominate the covers of the 1980s, including an oil spill, air and water pollution, toxic waste, and radioactive contamination from both nuclear power plants and nuclear weapons. In the 2000s, it is climate- and weather-related issues, particularly climate disruption, that dominate covers. Wildlife and habitat issues are covered steadily throughout the 1970s to the present. Resource plundering, overpopulation, and food issues are featured relatively infrequently.

Two related points need to be made here. First, the covers tend to focus on issues related to the unintended consequences of human activity (i.e., pollution, climate disruption, and species extinction) more so than on the actual human activities that bring them about (extraction, consumption, and waste). The fact that resource plundering and overpopulation get so little attention is significant because they represent key determinants of environmental degradation. Furthermore, another key determinant, the consumer lifestyle, is never questioned. On the contrary, many of *Time*'s covers (those outside the dataset) celebrate consumption and wealth accumulation. Thus, the covers in our set emphasize consequences rather than causes of environmental degradation.

The second point is that environmental issues are presented in a fragmented fashion far more often than in a holistic fashion. Very few covers are about the environment in general and few connect issues to causes, let alone to each other. In our view, this suggests that *Time*'s representations frequently overlook a key element needed to address environmental issues, namely, the underlying causes. In other words, whatever the text inside may be saying, the covers are not telling readers what's at the root of the problem.

Representations of Issues

Table 2 summarizes how these issues are represented visually. The most common ways the issues are represented is with animals, sources of pollution, and humans. Animals may be portrayed as suffering (e.g., oil-soaked bird) or dead (e.g., skeleton fish). However, the majority of animals used to represent issues are unharmed examples of the species at risk. Other relatively common ways of representing the issues include showing the sources of pollution (mushroom clouds, nuclear reactors, etc.), and suffering humans. In addition to effects on animals and humans, various ecosystem effects were also shown, including smog, oil on the water, and shrunken ice. In 10 instances, *Time* used visual metaphorical representations of the issues. They also used graphical design techniques to evoke negative conditions. In should also be noted that almost all of the 67 covers used text to represent the issues. In fact, some covers had only text representations.

Several points can be made about these representations. To begin with, the high frequency of textual representations of the issues suggests that text is necessary to anchor the meaning of the images in most cases. It is possible, then, to argue that adequately representing environmental issues visually is a challenge in many cases. For example, while the image of a mushroom cloud is itself powerful enough for people to get its denotative and connotative meaning, that is not the case with many other issues, most notably climate disruption.

Given *Time*'s history of featuring people on its covers, we were surprised to find that people were used infrequently to represent the issues. Although we coded seven instances of the use of suffering humans, only one really stands out affectingly on the covers, the one about the Union Carbide plant disaster in Bhopal (1984-12-17), which shows a close-up of a survivor who is clearly in anguish. We think the relative absence of suffering humans (let alone dead ones) is likely due to the fact that such images would be deemed too graphic for most magazine covers. On the other hand, we would not want to see suffering humans presented as the only symbol of these issues because that would serve to reinforce an anthropocentric view.

It also seems that showing dead or suffering animals is not something that editors want to feature on their covers. Thus, showing the impact on endangered species or simply animal victims is a challenge. Consider, for example, that we coded 12 instances where covers used unharmed, healthy-looking exemplars of species at risk to illustrate the issues (frequently species in decline). The species in those images, including sharks, tigers, a polar bear, a spotted owl, etc., were all portrayed as alive and well. But this imagery is in contrast to the textual claims that they were endangered. Here, the visuals seem to undermine the claim. On the other hand, an argument can be made for showing vibrant animals in order to get readers to appreciate them more.

More surprising was the fact that there were very few images of effects on ecosystems and many of those did not immediately signal an undesirable condition. The same sort of thinking may be at work here, the logic being that by showing a pristine landscape while claiming it is threatened will inspire concern.

In the previous section, we suggested that the covers emphasized consequences more than causes. We can return to this question when looking at how the issues are represented and here we find a slightly different picture. Looking just at the humans, animals, source of pollution, and ecosystem effects categories of representations, roughly half use the sources and roughly half use the consequences to represent the issues. In general, this is encouraging because it brings the viewer's attention closer to the causes of the degradation. However, when one looks at those sources, mushroom clouds from nuclear bombs represent almost half of them.

We noted 10 instances where visual metaphors were used to represent the issues. Some of these came off as clearly more effective than others. For example, the anthropomorphic pollution monster on a cover about oceans (1988-08-01) effectively conveyed the threat to ocean life and reinforced the tag line "Our Filthy Seas." On the other hand, the image of the earth as a fried egg in a cast-iron frying pan on the cover of an issue about "Global Warming" (2011-04-09) seems, at best, ambiguous.

Finally, the limited use of graphic design elements, such as colors, to convey meaning has, we think, unrealized potential. For example, a cover about species extinction (2009-04-13) shows a tiger whose face is fading into white nothingness where the tag line reads "Vanishing Act."

Overall, we find that even given the constraints of good taste in cover imagery required of a magazine such as *Time*, environmental issues could be more effectively represented with more powerful, evocative, and coherent visuals.

Aspects of Nature Represented

As Table 3 shows, four categories of aspects of nature account for most of the representations on these covers: animals, plants, places, and weather events. In addition, one cover featured a rock (shale) (2011-04-11). Animals and places dominate these representations of nature. The most commonly portrayed animals include birds ($n=15$), fish ($n=12$), dogs ($n=9$), and horses ($n=9$). The most commonly featured places are mountains ($n=17$), farms ($n=9$), oceans ($n=9$), forests ($n=7$), rivers ($n=7$), and the whole Earth ($n=7$). The most commonly featured plants were trees ($n=17$) and shrubs ($n=10$). The most common weather events were floods ($n=3$) and hurricanes ($n=3$).

On some covers, the aspects of nature are secondary to the main image, but in many others, they are foremost. In those cases, for example, we get close-ups of big cats, sharks, chimps, dogs, etc., all instantly recognizable animals for a mainstream audience.

Representations of Nature

Table 4 presents the frequencies of different types of representations of nature. Most prominent among these is nature as a resource, an idea that is consistently present throughout the decades. Nature as victim/patient and nature as problem are approximately equally present. Nature as victim/patient is a representation that emerges with the environmental movement in the 1970s. Nature as a research subject, human analog, or companion, and exemplar of fitness appear comparatively less frequently. The representation of nature as just itself was rare. These were cases where the aspect of nature was not assigned any particular relationship to humanity. In other words, it was not portrayed as either a resource, a problem, a companion/analog, a victim/patient, a research subject, or an exemplar of fitness.

The category of nature as a resource included everything from animal tools, like horses, to recreational places, like ski hills. It also included what we typically think of as resources: timber stands, farm animals, energy sources, etc. But in those cases, their extraction is rarely illustrated, but rather referenced in text. As discussed earlier, the using up of aspects of nature to feed consumption is not really referenced visually on these covers. Mostly what we get are food, entertainment, and recreational resources.

The images of nature as a problem are mostly comprised of floods and other unwanted weather events, as well as dangerous animals like sharks. When aspects of nature are represented as victims or patients, it is often done so with text rather than imagery, though there are some dead fish, endangered Amazon jungle animals, an oil-soaked bird and a few other images. But for the most part, there was nothing visually arresting in this category.

Table 3 Types of aspects of nature on *Time* covers by decade.

Aspects of nature	1920s	1930s	1940s	1950s	1960s	1970s	1980s	1990s	2000s	2010s[a]	Total
Covers with one or more aspects of nature[b]	3	7	3	11	6	12	18	16	26	5	107
Types of animals present (antelope, bear, bird, buffalo, butterfly, camel, capybara, cat, chicken, chimpanzee, coral, cow, deer, dog, dolphin, elephant, fish, fox, giraffe, gorilla, horse, insect, jaguar, lion, meat, moose, pig, polar bear, seal, shark, sheep, snake, tapir, tiger, and walrus)	3	8	9	11	8	16	18	10	12	3	98
Types of plants present (flowers, fruits & vegetables, shrubs, and trees)				4	1	7	8	4	8	1	33
Types of places present (beach, coastal cliff, desert, farm, forest, glacier, ice flow, jungle, lake, lawn, mountains, ocean, river, volcano, waterfall, and whole earth)			1	12	9	13	17	11	15	1	79
Types of weather events present (flood, hurricane, ice, lightning, rain, snow, sun, tornado, and wildfire)				3	3	5		3	6		20

[a]These numbers are for only two years of this decade.
[b]Some covers contain multiple aspects of nature, so simply totaling the other rows would give an inflated picture of the number of covers involved.

Table 4 Categories of representations of nature on *Time* covers by decade.

Representations of nature	1920s	1930s	1940s	1950s	1960s	1970s	1980s	1990s	2000s	2010s[a]	Total
Covers with one or more representations of nature[b]	3	7	3	11	6	12	18	16	26	5	107
Nature as resource	2	6	3	7	1	5	6	3	6	2	41
Nature as victim/patient						1	10	6	8	2	27
Nature as problem					1	1	6	3	3	7	21
Nature as research subject			1	3				4	5	1	14
Nature as human analog and/or companion	1	1			1	3	1	1	3	1	12
Nature as exemplar of fitness						2	1	4	3	1	11
Nature as just itself				1			1		2	3	7

[a]These numbers are for only two years of this decade.
[b]Some covers contain multiple representations of nature, so simply totaling the other rows would give an inflated picture of the number of covers involved.

A number of covers featured questions about how much other animals are like humans: can they think, what do they think, how did we evolve from them? These were coded nature as research subject. There were also a number of covers featuring companion animals, including some wild animals. These were coded as companion/analog. In some cases, both of these codes were applicable to the same covers.

Finally, a few of the representations did not situate the aspects of nature in relation to humans, and they were coded as nature as just itself. These are important to acknowledge because, in some ways, they are acknowledgments of the independence of non-human nature from human will.

Overall, these are typical representations found in the mainstream media which echo Meisner's earlier findings (2005). The emphasis on nature as a resource and nature as a problem can be seen as contradicting the representations of nature as a victim or patient in need of our care or as just itself. Or, they can be read together as a kind of symbolic domestication of nature (Meisner, 2005).

People

We did not include a table showing the different types of people featured on these covers. However, we can summarize its key results. Citizens were the most common type of person featured ($n=21$). Following that, politicians ($n=10$) and scientists ($n=8$) were also reasonably prominent. The scientists included meteorologists and other researchers. Only one person was represented as an environmentalist/ecologist on these covers, Barry Commoner (1970-02-02). Jacques Cousteau also appears, but he is not presented as an environmentalist, nor does the cover he is on (1960-03-28) feature an environmental issue; it does feature aspects of nature. One wonders why people like Rachel Carson, Paul Ehrlich, and James Lovelock never made *Time*'s cover. Interestingly, Ralph Nader (1969-12-12) and Al Gore (2007-05-28) each had their own covers in our original pool, but like Cousteau, were not in any way represented as environmentalists. Incidentally, Gore appears on 12 covers of *Time*, but nowhere are his environmental credentials mentioned.

Generally speaking, *Time* covers frequently featured men's faces from the 1920s to the 1950s. These were people who had done things, and who had agency. But from the 1960s onward, people are much less prominent on the covers, and when they do appear, they are frequently victims of the environmental issue. After the Barry Commoner cover in 1970, there are almost no agents of environmental change portrayed. The one notable exception to this is a "special environment issue" on "how to win the war on global warming" (2008-04-22), featuring soldiers raising a redwood tree in the vein of the flag raising on Iwo Jima. This is a metaphor, of course. The absence of environmental change agents on the covers is even more perplexing, given the fact that the magazine has, on a number of occasions, featured sections on "Heroes of the Environment" or "Heroes for the Planet."

Actions

We found that very few actions to address environmental degradation were featured on these covers. Furthermore, most that did were represented only by text. The few exceptions to this included a cover on green design (2009-05-03) featuring an intricate green leaf and bird pattern, a cover on energy efficiency (2009-01-12) featuring a compact fluorescent light bulb in place of a man's head, and the redwood-raising Iwo Jima issue (2008-04-22). The cover featuring Barry Commoner (1970-02-02) also has a partly aspirational cover in that the left side shows an example of a healthy and verdant landscape, even if it does not suggest any specific actions for how to make that happen. Other hopeful (if not exactly action-oriented) covers come from two special issues (2002-08-26 and 2000-04-26) featuring the whole Earth metaphorically represented as a flower and a butterfly, respectively.

The fact that so few covers offer representations of actions that can be taken to address environmental degradation is not altogether unexpected. Much environmental discourse has focused on issues and problems in the past. It is only more recently, with the shift to a focus on "sustainability," that we are seeing more solution-oriented representations. But it is also disturbing that the kinds of change-agents and positive actions that are described within the magazine are not making it onto the covers.

Conclusions

In this paper, we have presented a first longitudinal study of the ways in which a leading news magazine, *Time*, has visually represented aspects of nature and environmental affairs on its covers. We have suggested that magazines such as *Time* can be important cultural artifacts that reflect societal trends and that provide visual cues that can be used by citizens to better understand the issues represented.

In this preliminary examination, we have considered both quantitative and qualitative aspects of the cover imagery, with an emphasis on the former. From this, we draw several conclusions about the patterns of representation in our dataset.

First, the small percentage of covers found in this survey suggests that in *Time*'s larger scheme, environmental affairs are not a high priority. The numbers, though, do appear consistent with the results reported by McGeachy (1989) in her study of the coverage of environmental issues in six magazines including *Time*. In general-oriented magazines, 0.9% of the pages were devoted to environmental issues (McGeachy, 1989).

However, despite the small percentage of covers focusing on environmental affairs overall, the content analysis shows a subtle trend toward the increasing use of nature and environmental issues on the covers of *Time*. This trend is consistent with those found in other media, and is also consistent with the idea of the "mainstreaming" of environmental issues, something Loi (2010) found in the case of climate disruption. Of course, how issues are covered matters as well.

With respect to how the magazine visually addresses issues, we conclude that *Time* takes a position against pollution, climate disruption, species extinction, and habitat destruction on its covers. However, it does not appear to question the resource extraction and consumption that drives those downstream consequences with any significant visual force on the covers. The focus on consequences more than causes suggests a superficial outlook with respect to environmental degradation and draws attention away from the deeper roots of the problem. Furthermore, the pattern of attention to issues reinforces the fragmented approach to environmental affairs of most media coverage.

This is supported by a pattern of representations of nature that suggest at best a contradictory attitude toward the nonhuman world. In some cases, *Time*'s covers evoke concern and care toward other nature, but more often they echo the dominant anthropocentric-resourcist ideology.

In terms of how the issues are portrayed and the symbols are used in these representations, we conclude that *Time*'s covers could be more powerful, emotionally engaging, and coherent. Much of the imagery lacks affective force, or worse, contradicts the claims being made. In this respect, the covers often seem like a missed opportunity to present meaningful, memorable, and provocative images. That said, since this is not a comparative study, we do not want to imply that *Time* is unique in this regard. So we are not saying that *Time* is any better or worse than its competition.

Finally, the covers almost completely neglect to represent environmental leaders and role models and have only recently begun to present possible courses of action on environmental issues. We conclude that overall, *Time*'s covers take a shallow position on environmental affairs, argue it weakly, let slip the opportunity to use powerful imagery, and offer few solutions and little agency.

We acknowledge the commercial pressures under which the magazine must operate, but our view is that given its advocacy stance, the magazine could do better in representing environmental affairs in such a way as to drive them higher up on the agenda of the public and the policy makers.

References

Allan, S., Adam, B., & Carter, C. (Eds.). (2000). *Environmental risks and the media*. New York, NY: Routledge.
Anderson, A. (1997). *Media, culture and the environment*. New Brunswick, NJ: Rutgers University Press.
Angeletti, N., & Oliva, A. (2010). *Time: The illustrated history of the world's most influential magazine*. New York, NY: Rizzoli.
Baker, S. (1993). *Picturing the beast: Animals, identity and representation*. Manchester: Manchester University Press.
Bates, S. (2011). Public intellectuals on *Time*'s covers. *Journalism History, 37*(1), 39–50.
Bell, P. (2001). Content analysis of visual images. In T. Van Leeuwen & C. Jewitt (Eds.), *Handbook of visual analysis* (pp. 10–34). London: Sage.
Best, J. (1995). Typification and social problems construction. In J. Best (Ed.), *Images of issues: Typifying contemporary social problems* (2nd ed., pp. 3–10). New York, NY: Aldine de Gruyter.
Bousé, D. (2000). *Wildlife films*. Philadelphia, PA: University of Pennsylvania Press.
Boykoff, M. T. (2011). *Who speaks for the climate? Making sense of media reporting on climate change*. Cambridge: Cambridge University Press.
Brereton, P. (2004). *Hollywood Utopia: Ecology in contemporary American cinema*. Bristol: Intellect.
Cerulo, K. A. (1984). Television, magazine covers, and the shared symbolic environment: 1948–1970. *American Sociological Review, 49*(4), 566–570. doi:10.2307/2095469
Christ, W. G., & Johnson, S. (1985). Images through *Time*: Man of the Year covers. *Journalism Quarterly, 62*(4), 891–893. doi:10.1177/107769908506200428
Corbett, J. B. (2006). *Communicating nature: How we create and understand environmental messages*. Washington, DC: Island Press.
Dale, S. (1996). *McLuhan's children: The Greenpeace message and the media*. Toronto, ON: Between The Lines.
DeLuca, K. (1999). *Image politics: The new rhetoric of environmental activism*. New York, NY: Guilford Press.
DiFrancesco, D. A., & Young, N. (2011). Seeing climate change: The visual construction of global warming in Canadian national print media. *Cultural Geographies, 18*(4), 517–536. doi:10.1177/1474474010382072
Doyle, J. (2007). Picturing the clima(c)tic: Greenpeace and the representational politics of climate change communication. *Science as Culture, 16*(2), 129–150. doi:10.1080/09505430701368938
Doyle, J. (2011). *Mediating climate change*. Burlington, VT: Ashgate.
Evernden, N. (1992). *The social creation of nature*. Baltimore: Johns Hopkins.
Hannigan, J. A. (1995). *Environmental sociology: A social constructionist perspective*. New York, NY: Routledge.
Hansen, A. (Ed.). (1993). *The mass media and environmental issues*. Leicester: Leicester University Press.
Hansen, A. (2010). *Environment, media and communication*. New York, NY: Routledge.
Hansen, A. (2011). Communication, media and environment: Towards reconnecting research on the production, content and social implications of environmental communication. *International Communication Gazette, 73*(1–2), 7–25. doi:10.1177/1748048510386739
Hansen, A., Cottle, S., Negrine, R., & Newbold, C. (1998). *Mass communication research methods*. New York, NY: New York University Press.
Hansen, A., & Machin, D. (2008). Visually branding the environment: Climate change as a marketing opportunity. *Discourse Studies, 10*(6), 777–794. doi:10.1177/1461445608098200
Howenstine, E. (1987). Environmental reporting: Shift from 1970 to 1982. *Journalism Quarterly, 64*(4), 842–846. doi:10.1177/107769908706400425
Ingram, D. (2000). *Green screen: Environmentalism and Hollywood cinema*. Exeter: University of Exeter Press.

Ivakhiv, A. (2010). Nature's nation: Improvisation, democracy, and Ken Burns' national parks. *Environmental Communication: A Journal of Nature and Culture, 4*(4), 462–468. doi:10.1080/17524032.2010.522141

Johnson, S. (2002). The art and science of magazine cover research. *Journal of Magazine & New Media Research, 5*(1), 1–10.

Johnson, S., & Christ, W. G. (1988). Women through *Time*: Who gets covered? *Journalism Quarterly, 65*(4), 889–897. doi:10.1177/107769908806500408

Kalof, L., & Fitzgerald, A. (2003). Reading the trophy: Exploring the display of dead animals in hunting magazines. *Visual Studies, 18*(2), 112–122. doi:10.1080/14725860310001631985

Lawrence, E. A. (1993). The sacred bee, the filthy pig, and the bat out of hell: Animal symbolism as cognitive biophilia. In S. R. Kellert & E. O. Wilson (Eds.), *The biophilia hypothesis* (pp. 301–341). Washington, DC: Island Press.

Leath, V. M., & Lumpkin, A. (1992). An analysis of sportswomen on the covers and in the feature articles of women's sports and fitness magazine, 1975–1989. *Journal of Sport & Social Issues, 16*(2), 121–126. doi:10.1177/019372359201600207

Leiss, W., Kline, S., & Jhally, S. (1990). *Social communication in advertising: Persons, products and images of well being* (2nd ed.). Toronto, ON: Nelson Canada.

Lester, L. (2010). *Media and environment: Conflict, politics and the news*. Cambridge: Polity.

Lester, L., & Cottle, S. (2009). Visualizing climate change: Television news and ecological citizenship. *International Journal of Communication, 3*, 920–936.

Lindahl, E. N. (2006). *Mediating nature: Environmentalism and modern culture*. New York, NY: Routledge. Retrieved from http://ijoc.org/ojs/index.php/ijoc/article/view/509

Linden, E. (2006). *The winds of change: Climate, weather, and the destruction of civilizations*. New York, NY: Simon & Schuster.

Loi, R. (2010). Case study: Climate change reporting in *Time* magazine. In: C. Lever-Tracy (Ed.), *Routledge handbook of climate change and society* (pp. 219–229). New York.

Manzo, K. (2010a). Imaging vulnerability: The iconography of climate change. *Area, 42*(1), 96–107. doi:10.1111/j.1475-4762.2009.00887.x

Manzo, K. (2010b). Beyond polar bears? Re-envisioning climate change. *Meteorological Applications, 17*(2), 196–208. doi:10.1002/met.193

Matsa, K., Sasseen, J., & Mitchell, A. (2012). Magazines: Are hopes for tablets overdone? In *2012 State of the News Media*. The Pew Research Center's Project for Excellence in Journalism. Retrieved from http://stateofthemedia.org/2012/magazines-are-hopes-for-tablets-overdone/

McGeachy, L. (1989). Trends in magazine coverage of environmental issues, 1961–1986. *The Journal of Environmental Education, 20*(2), 6–13. doi:10.1080/00958964.1989.9943025

McQuail, D. (2005). *McQuail's mass communication theory* (5th ed.). Thousand Oaks: Sage Publications.

Meisner, M. (2005). Knowing nature through the media: An examination of mainstream print and television representations of the non-human world. In G. B. Walker & W. J. Kinsella (Eds.), *Finding our way(s) in environmental communication: Proceedings of the seventh biennial conference on communication and the environment* (pp. 425–437). Corvallis: Department of Speech Communication, Oregon State University.

Meisner, M. S. (1997). Pitching the beast: Representations of non-human animals in contemporary print advertising. In S. Senecah (Ed.), *Proceedings of the fourth biennial conference on communication and environment* (pp. 302–314). Syracuse: State University of New York College of Environmental Science and Forestry.

Mitman, G. (1999). *Reel nature: America's romance with wildlife on film*. Cambridge: Harvard University Press.

Moore, W. S., Bowers, D. R., & Granovsky, T. A. (1982). What are magazine articles telling us about insects? *Journalism Quarterly, 59*(3), 464–467. doi:10.1177/107769908205900318

Moriarty, S., & Popovich, M. (1991). Newsmagazine visuals and the 1988 Presidential election. *Journalism Quarterly, 68*(3), 371–380. doi:10.1177/107769909106800307

Nicholson-Cole, S. A. (2005). Representing climate change futures: A critique on the use of images for visual communication. *Computers Environment and Urban Systems, 29*(3), 255–273. doi:10.1016/j.compenvurbsys.2004.05.002

Peeples, J. (2011). Toxic sublime: Imaging contaminated landscapes. *Environmental Communication: A Journal of Nature and Culture, 5*(4), 373–392. doi:10.1080/17524032.2011.616516

Perlmutter, D. D. (2007). *Picturing China in the American press: The visual portrayal of Sino–American relations in Time magazine 1949–1973*. Lanham, MD: Lexington Books.

Pompper, D., Lee, S., & Lerner, S. (2009). Gauging outcomes of the 1960s social equality movements: Nearly four decades of gender and ethnicity on the cover of the *Rolling Stone magazine*. *The Journal of Popular Culture, 42*(2), 273–290. doi:10.1111/j.1540-5931.2009.00679.x

Remillard, C. (2011). Picturing environmental risk: The Canadian oil sands and the national geographic. *International Communication Gazette, 73*, 127–143. doi:10.1177/1748048510386745

Schoenfeld, A. C. (1983). The environmental movement as reflected in the American magazine. *Journalism Quarterly, 60*(3), 470–475. doi:10.1177/107769908306000312

Scott, D. W., & Stout, D. A. (2006). Religion on TIME: Personal spiritual quests and religious institutions on the cover of a popular news magazine. *Journal of Magazine & New Media Research, 8*(1), 1–17.

Seppänen, J., & Väliverronen, E. (2003). Visualizing biodiversity: The role of photographs in environmental discourse. *Science as Culture, 12*(1), 59–85. doi:10.1080/0950543032000062263

Shanahan, J., & McComas, K. (1999). *Nature stories: Depictions of the environment and their effects.* Cresskill, NJ: Hampton Press.

Slawter, L. D. (2008). TreeHuggerTV: Re-visualizing environmental activism in the post-network era. *Environmental Communication: A Journal of Nature and Culture, 2*(2), 212–228. doi:10.1080/17524030802141760

Smith, N. W., & Joffe, H. (2009). Climate change in the British press: The role of the visual. *Journal of Risk Research, 12*(5), 647–663. doi:10.1080/13669870802586512

Soper, K. (1995). *What is nature? Culture, politics and the non-human.* Oxford: Blackwell.

Stengel, R. (2010). Inside the red border. *Time.com.* Retrieved from http://www.time.com/time/magazine/article/0,9171,1997446,00.html.

Sumner, D. E. (2002). Sixty-four years of *Life*: What did its 2,128 covers cover? *Journal of Magazine & New Media Research, 5*(1).

Todd, A. M. (2010). Anthropocentric distance in National Geographic's environmental aesthetic. *Environmental Communication: A Journal of Nature and Culture, 4*(2), 206–224. doi:10.1080/17524030903522371

Waldman, P., & Devitt, J. (1998). Newspaper photographs and the 1996 Presidential election: The question of bias. *Journalism & Mass Communication Quarterly, 75*(2), 302–311. doi:10.1177/107769909807500206

Wilson, A. (1991). *The culture of nature: North American landscape from Disney to the Exxon Valdez.* Toronto: Between The Lines.

Appendix. List of issues in the dataset

The following 128 covers were in our dataset:

1926-07-12, 1928-02-27, 1929-03-18, 1930-03-03, 1932-04-18, 1934-08-20, 1936-06-08, 1937-03-29, 1937-05-10, 1938-02-21, 1941-07-21, 1946-04-29, 1946-07-01, 1947-07-07, 1950-08-28, 1950-10-09, 1954-04-12, 1954-05-31, 1954-08-23, 1955-01-17, 1955-10-24, 1956-09-03, 1956-12-17, 1956-12-31, 1959-02-09, 1959-08-17, 1960-01-11, 1960-03-28, 1961-07-14, 1962-07-20, 1963-04-05, 1964-09-25, 1965-10-01, 1967-01-27, 1969-08-08, 1970-02-02, 1970-07-27, 1973-04-09, 1973-12-03, 1974-12-23, 1975-06-23, 1976-07-12, 1977-01-31, 1977-12-12, 1978-06-05, 1978-11-06, 1979-04-09, 1979-12-24, 1980-01-21, 1980-06-02, 1980-09-22, 1980-12-15, 1981-12-07, 1982-03-29, 1984-08-06, 1984-12-17, 1985-02-18, 1985-07-29, 1985-10-14, 1986-05-12, 1987-02-23, 1987-08-10, 1987-10-19, 1988-07-04, 1988-08-01, 1988-10-31, 1989-01-02, 1989-03-27, 1989-04-17, 1989-07-24, 1989-09-18, 1989-10-16, 1990-01-15, 1990-06-25, 1991-04-29, 1991-07-22, 1991-08-05, 1991-10-07, 1991-11-18, 1992-02-17, 1992-06-01, 1992-07-13, 1992-08-10, 1993-03-22, 1993-07-26, 1994-03-28, 1994-12-12, 1995-08-14, 1995-09-04, 1996-03-04, 1996-05-20, 1997-03-10, 1997-08-11, 1998-11-23, 2000-04-26, 2000-07-31, 2001-04-09, 2001-07-16, 2001-07-23, 2001-07-30, 2002-08-26, 2004-08-23, 2005-08-01, 2005-08-15, 2005-09-12, 2005-10-03, 2005-11-28, 2006-04-03, 2006-10-09, 2007-03-12, 2007-04-09, 2007-08-13, 2007-10-01, 2007-11-05, 2008-03-24, 2008-04-07, 2008-04-22, 2009-01-12, 2009-03-23, 2009-04-13, 2009-05-03, 2009-08-31, 2010-05-17, 2010-06-21, 2010-08-16, 2011-03-28, 2011-04-11, 2011-07-18.

Sporting Nature(s): Wildness, the Primitive, and Naturalizing Imagery in MMA and Sports Advertisements

Matthew P. Ferrari

This essay examines two Nike commercials, a TapOut commercial, and the proliferation of mixed martial arts (MMA) t-shirt visual culture, all of which symbolically link wildness and "nature" as primitivity to their particular sport contexts. MMA in particular, it is argued, is more symbolically available to symbolic discourses of the "natural" and the "primitive" because of the sport's technological minimalism. The MMA t-shirt is posited as a safe, masculine primitive performance, functioning as an expressive personal substitute (or supplement) to the analogous tattoos "worn" by many fighters and fans. Additionally, this paper reviews and connects several disparate bodies of literature, moving from a discussion of eco-critical principles for critiquing the cultural production of nature / the natural, to an assessment of "nature" as primitivity, and finally to highlight how critical analyses of sport and MMA implicate related categories. While environmental communication has addressed the place of "nature" in advertising, little has been written about how discourses of nature, gender, and the environment intersect with the highly mediatized culture of sports. This article adds to the subfield by initiating just such a critical discussion. Finally, I contend that one of the main ideological functions of the employment of nature imagery here is to implicitly authorize notions of wildness or the "primitive" in close association with a male animal ideology, and also to symbolically reinforce existing narratives which naturalize aggression. These advertisements posit, I argue, a metaphysical rather than realist ecological discourse, enabling an unsustainable narrative of the naturalness of human-on-human violence and aggression.

Matthew P. Ferrari is a Ph.D. Candidate in the Department of Communication, University of Massachusetts-Amherst.

Cage Fighting, or Mixed Martial Arts (MMA) as it has come to be known officially, is by some accounts the fastest growing spectator sport in the world. It is also, as I interrogate here, an emergent cultural scene with a heightened availability to deployments of ideologically constructed nature/naturalizing imagery, raising issues highly pertinent to the fields of environmental communication (EC), green cultural studies (GCS), the sociology of sport, and the study of social and sexual power relations. While also present in sport more generally, within the cultural arena of MMA, iconographies of "nature" have established a qualitatively different form of visibility than in other sports. One sees t-shirts, logos, and multi-media with fragmentary plant and animal features, such as wings, thorns, fangs, eyes, claws, and horns. When "nature" is deployed in visual forms in order to generate associations between masculine-centered combat culture and notions of the "natural," violence and aggression are symbolically authorized, or naturalized. The cultural concepts of "nature" and the "primitive" on display in MMA (and sport in general) are an under-examined site in which narratives of acceptable human behavior are put forth; where the symbolic meanings of images are incorporated into value systems, and worldviews are directly affected.

Consider several examples. Frankie Edgar, a popular UFC (The Ultimate Fighting Championship) lightweight, enters the arena for UFC 112 in Abu Dhabi wearing an Affliction brand t-shirt featuring elaborate blue wings extending from his chest over his shoulder and down to his mid-back (See Figure 1). The wings are arranged on the

Figure 1. (Color online) Selected stills from UFC 112, highlighting Edgar's "Affliction" walk-out t-shirt (author-produced montage).

shirt roughly corresponding to the wearer's physiology—where wings might fall on a human—suggesting a theriomorphic (i.e. human-to-animal, rather than anthropomorphic, animal-to-human) metamorphosis. In the image, depicting Edgar reacting just moments after beating B.J. Penn for the lightweight title, we see the t-shirt, and below this on his shorts is a Venum brand logo featuring the stylized figure of a snake with exposed fangs. As another example, consider the Strikeforce logo (a smaller promotion owned by the UFC), configured graphically as two outstretched wings flanking a large "S", and two gloved hands protruding under the brand name, with blood splatter anchoring the background. This logo would be displayed in television advertisements, worn on clothing, and emblazoned at different locations during events. A cursory internet search for MMA logos and apparel, or at a site like www.fighterstyle.com, reveal this profusion of fragmented nature imagery. And while nature symbols are certainly not the only type of imagery on display here, their visibility through repetition and ideological function(s) warrants further examination.

The pattern within this discursive dimension of MMA's visual culture is for nature imagery to be represented in fragments rather than wholes, depicted as central or framing visual motifs within the graphic design of logos, apparel, and advertisements. Quite significantly, nature images in this context are less frequently depicted with photographic verisimilitude, but are instead highly stylized, formalized, or generic. In these instances animal and plant imagery stand in not as individuals, but as "species representatives," functioning as "shorthand symbols" for human values (Corbett, 2006, p. 207; Hansen, 2010, p.138). The iconic elements of animals (and plants to a lesser extent) signify defense and aggression mechanisms in the natural world, combined in expressive ways with other graphic elements, and of course, a brand name. And while symbolic constructions of nature are heightened in MMA, I also wish to draw a connection here between the symbolic uses of nature and the promotional imaginary of mainstream commercial sports in general. Several recent commercials by Nike—a company whose ads are so often conceptually organized through a romantic impulse towards essentializing views of human/athlete performance—are examined to articulate and further bear out this cultural discourse. "Alter Ego" and "For Warriors Only," promoting Nike's line of "Pro Combat Apparel," stand out as two of the most striking visual articulations of a sports marketing imaginary that symbolically merges the categories of human and animal, culture and nature.

Through the analysis of a thematically similar grouping of MMA t-shirts and several television advertisements, I consider the key ideological features and functions of sports' availability to symbolic constructions of nature, wildness, and primitivity. More specifically, I examine here the following: 1) The phenomenon of the MMA "walk-out t-shirt" and its common use of fragmented nature imagery emphasizing predation, and thereby, the significance of (literally) wearing—or "sporting," if you will—these sorts of nature symbols within an MMA context. 2) A recent television commercial for Tap Out brand MMA apparel depicting a fighter with digitally incorporated animal features. 3) The use of visual analogies between plant, animal, and athlete in two television commercials for Nike's "Pro Combat" apparel. Through these examples, I wish to demonstrate that the promotional imaginary of sport in general is an important cultural arena of symbolic availability for the flexible deployment of symbolic "nature(s)."

Overall, my interest here is with the visual symbols—and in particular the use of fragmented nature imagery—which I argue function to authorize notions of wildness, literally and figuratively "natural-izing" (i.e. biologically essentializing, rendering timeless and ahistorical) masculine aggression and violence through recuperative/regenerative uses of the "primitive" (Torgovnick, 1990, 1997; White, 1985). These symbolic uses of nature are configured "as primitivity" (Soper, 1995), and in close association with a historically troubled "male animal ideology" (Bordo, 1999), or a revived version of the "masculine primitive" (Bederman, 1995; Rotundo, 1993). I view these cultural constructions of nature as functioning symbolically to reinforce the dominant social order's existing naturalization(s) of aggression, within what might be thought of as the pervasive and perpetually "violent habitus of Western modernity" (Jantzen, 2002, p.8).

Critiquing the Cultural Production of Nature/Naturalization(s)

The field of environmental communication (EC), and what has been labeled green cultural studies (GCS) (Hochman, 1998), provide an important theoretical forum for interrogating the symbolic uses of nature in various cultural contexts. Scholars working in EC have taken up questions of how nature imagery is deployed in adaptable and often ideologically contradictory ways in advertising discourse, especially as it relates to ideologies of environmental exploitation, cultural and national identity, and issues of race and gender (e.g. Corbett, 2006; Cox, 2010; Hansen, 2010; Meister & Japp, 2002). In explaining why images of nature—and our cultural notions of what is "natural"—are so frequently exploited in advertising and popular culture discourses, Anders Hansen explains that associating things with the natural world "serves to hide what are essentially partisan arguments and interests and to invest them with moral or universal authority and legitimacy" (2010, p.136). That is, nature is constructed or rhetorically invented according to human needs to "justify social patterns," market imperatives, and perhaps most crucially, in order to disguise or conceal this very constructedness (Hansen, 2010, p.137). In this way, Kate Soper argues, "representations of nature, and the concepts we bring to it, can have very definite political effects" (1995, p.9). In particular, she emphasizes effects such as the naturalization of social and sexual power relations, which are especially relevant to the analysis at hand. Furthermore, Soper posits that "'nature' in human affairs is a concept through which social conventions and cultural norms," such as violence and aggression, for example, "are continuously legitimated and contested" (p. 33).

EC and GCS overlap through their shared goals of understanding the processes of communicating nature and the environment. I turn to GCS especially for its vicinity to cultural studies' theoretical legacy of successfully deconstructing the category of the "natural." The legacy of Stuart Hall, Raymond Williams, and other Frankfurt and

Birmingham school theorists paved the way for cultural studies' fundamental critiques of capitalist conditions of "reification, or naturalization, that situation where culture replaces nature as the realm of the given, the unchangeable" (Hochman, 1998, p. 6). GCS is organized around critiquing representations of nature. Ultimately the charge of GCS is, according to Hochman, "the examination of proliferating cultural representations of nature—i.e., lexical, pictorial, and actual manipulations of plants, animals, and elements—for their potential to affect audiences affecting nature-out-there" (1998, p. 8).

Environmental communication scholars have made important observations on the psychology of advertising's appeals through "nature." Drawing on the work of Pollay (1986) and Lasch (1978), Corbett (2006) explains that successful advertising acts upon personal dissatisfactions and insecurities, not necessarily one's actual needs. The most common form of nature imagery used in advertisements, Corbett notes, is "nature as backdrop" (2006, pp. 149–150). In this formal arrangement nature is less likely to be appreciated in itself, but functions symbolically in a larger rhetorical action in which "advertisers are depending on qualities and features of the nonhuman world to help in selling the message" (2006, p. 150). For example, a Chevy Silverado SUV is presented in slow motion, tearing aggressively through a snow-covered mountaintop. The backdrop of a pristine natural environment lends seductive appeal to the truck as a "vehicle" empowering one to escape domestic confinement and experience the beauty of remote landscapes from a position of relative safety and comfort. "Nature" is not the thing for sale, but is merely a symbol giving the product greater meaning through association. In the crude sense, the image of "nature" is exploited to sell something that despoils nature. Interrogating and short-circuiting such cultural constructions of nature is a crucial task of GCS, as the symbolic domination of nature has a reciprocal relationship with its material domination, each form empowering and perpetuating the other (Hochman).

In order to properly critique the use of nature imagery, it is necessary to gain a handle on the patterns and tendencies of its deployments. While the uses change historically, it is possible to isolate key thematic and ideological tendencies. Hansen (2010) generated a useful list of groupings, highlighting, among other things—and of particular interest to this study—nature as "threat," and nature as "challenge/sport/manhood/endurance" (p.145). Here nature's value is in its properties that test people and things, often put to use in displaying manhood, stamina, physical prowess, or a product's durability and reliability (Hansen).

When nature is depicted within this thematic grouping, related to manhood, stamina, physical prowess, sport, and so on, it is often configured through nature's categorical association with the "primitive." Nature is represented in these instances in gendered ways, as projections for masculine identity and desire (Soper, 1995). The next section considers the notion of "nature as primitivity," and how a historical myth of the "primitive" has taken specific gendered forms. While the "primitive" and naturality has often been historically associated with femininity, it has also taken a particular character and coding as a masculine primitive. In general terms, Western thought has historically tended to equate woman with reproduction and thus nature, while man is equated with production and thus culture (Soper, 1995). At certain times, however, cultures of masculinity have seen fit to appropriate signs of nature, often as a defensive reaction against the perceived threats of feminization, alienating forms of labor, over-civilization, and other threats to hegemonic masculinity (Bederman, 1995; Bordo, 1999; Rotundo, 1993).

Nature as Primitivity, and the Masculine Primitive

One of the secondary analytical goals of this study is to establish a stronger critical link between representations of nature and socio-cultural discourses of the "primitive," highlighting their often overlooked or occluded imbrications, especially in the context of sport and MMA as they impinge on larger cultural narratives about the naturalness of masculine-based violence. "Nature" (and its key associations, like the "primitive") is a key category through which our human status is negotiated. Soper argues that "Western configurations of nature—notably its association with the "primitive," the "bestial," the "corporeal" and the "feminine"—reflect a history of ideas about membership of the human community and ideals of human nature, and thus function as a register or narrative of human self-projections" (1995, p. 10).

According to Hayden White, twentieth-century primitivism quite often functioned as a social critique of the "over-civilized" society's potential threat to, or diminishment of, masculinity. Primitivism responds to a desire to "escape the obligations laid upon us by involvement in current social enterprises" (White, 1985 p.170). "Nature" and the "primitive" are thus tools of social critique through which one may temporarily escape or transgress the confinement of social norms. Primitivism, White explains, "simply invites men to be themselves, to give vent to their original, natural, but subsequently repressed desires, to throw off the restraints of civilization and thereby enter a kingdom that is naturally theirs (...)" (1985, p. 171).

Contrary to this utopian-sounding ideal, the masculine primitive—or what Susan Bordo (1999) similarly defines as a "male animal ideology" and "primal masculinity"—creates a problematic "double bind" between wildness and civility that poses challenges for young men to live up to. The "double bind" of masculinity—that is, the contradictory messages and expectations we impose on boys telling them to be "animals" or "beasts" in the arena of sports and competition, and civilized gentleman outside those arenas—is associated with the enduring if mutable operation of "the male animal as ideology" (Bordo, 1999, p. 245). Bordo links this ideology directly to modern organized sporting traditions that function ideologically as a site of masculine regeneration and recovery. Male animal ideology is a direct response to anxiety over "the repressive effects of civilization and its softening of men," which responds with "fantasies of recovering an unspoiled, primitive masculinity," carrying with it "a flood of animal metaphors" (Bordo, 1999, p.249).

The myth of the "primitive" and its historical status as a discursive social construct has been the subject of much scholarly writing (Diamond, 1974; Fabian, 1983; Kurasawa, 2002; Lovejoy & Boas, 1965; Torgovnick, 1990, 1997; White, 1985).

In much the same way as cultural constructions of "nature" function to help us measure and police our status as "human," the "primitive," through various "tropes of wildness," especially expressions of "primitive alterity," functions similarly as "a shifting construct discursively produced by Euro-American human sciences over the course of the modern era" (Kurasawa, 2002, p. 2–3, 10). According to Kurasawa, primitiveness "represents a related set of beliefs and values generated to explain rhetorically what Euro-American societies have become in relation to their pasts and futures" (2002, p. 2). MMA and Sports marketing imaginaries, through their use of nature imagery and tropes of wildness, are profound popular culture examples of what Lovejoy and Boas describe as primitivism's "backwards looking habit of mind," an impulse towards "a recovery of what has been lost" in civilized, rational modernity (1965, p. 7). In a manner similar to the semiotic flexibility of "nature"—its potential to be constructed in wholly ambivalent or contradictory ideological modes—so too has the "primitive" been produced variously as a form of "symbolic domination through representational means," and as a myth employed "to radically put into question existing institutions, values, and habits" of society (Kurasawa, 2002, p 2–3).

It is important to highlight that the "primitive" as a discursive construct now typically refers to an "internalized" or "socio-psychological" primitivism, rather than an earlier "external" primitivism which was the problematic subject of early work in anthropology and turn of the century racist social-evolutionary discourses (Kurasawa, 2002; White, 1985). Thus it is necessary to make a distinction between the "primitive" as a powerful racist discourse of "evolutionary distancing"—one which Western civilization has historically used to define itself and justify colonial oppression—and a "socio-psychological primitivism" more closely associated with the alienated modern subject's romantic or nostalgic impulse to re-access their "natural" or untamed selves (Fabian, 1983; Torgovnick, 1990; White, 1985). Torgovnick argues that in the twentieth-century primitivism went from being a discourse used to justify the imperialistic subjugation of other populations, to a "medium of soul searching and self-transformation" for the modern Western subject (1997 p. 13).

Around the turn of the twentieth century, primitivism as a discourse of self-transformation took a very specific form in relation to masculinity, emphasizing the survival instincts, violence and impulsivity associated with "natural" or "primitive masculinity," and contrary to restrained Victorian manliness (Bederman, 1995). It is important to acknowledge that "nature" and the "primitive" have not always been associated with violence, savagery, and uncivilized behavior, having served at other times as discourses promoting benevolence, peace, and harmony with "nature." Yet for the purposes of this study, it is more important to understand a history of adapting "nature" as anti-civilization, as primitivity, to reinvent, reinscribe, and represent powerful manhood (Bederman, 1995).

Primitivism's impulse towards exploring deep connections between humans and animals is also relevant here. The "primitive," Torgovnick explains, "begins with the discontinuities separating human bodies, animals, and inanimate things—and seeks to bridge the gap" (1997, p. 7). A subset of primitivism described as "animalitarianism" views man as "a perverted creature profoundly alienated from nature," ennobling the animal by comparison as "more admirable" and possessed of greater "physical vigor and bodily endowments" (Lovejoy & Boas, 1965, p. 19–22).

Animals are claimed to have been the first subjects in painting, and possibly even the first metaphor (Berger, 2009). Berger claims that we are fundamentally drawn to looking at animals because of their place as "an intercession between man and his origins" (Berger, 2009, p. 253). Animals (as nature) serve as a central value concept, often configured symbolically in opposition to "the social institutions which strip man of his natural essence and imprison him" (Berger, 2009, p.257). Our relationship to animals is established and perpetuated through "master narratives" dating back to classical Western thought and literature. For instance, to men, animals are capable of embodying desirable qualities, such as power, strength, speed, bravery, and spirit (Magdoff & Barnett, 1989). Hence their frequent use in advertising discourses, such as selling cars, or as sports team mascots. The place of "nature" in our understanding of sports, or the "naturalness" of certain sports, becomes especially heightened in recent scholarly discussions of MMA, while existing less explicitly in literature on the history of sport in general.

Sport Studies' Uneasy Recourse to the "Primitive" and "Natural"

To be naturalized, Johannes Fabian explains, is "to be separated from (historic) events meaningful to mankind" (1983, p.13). Social history is by definition concerned with de-naturalizing socio-historical events and phenomena which become codified in popular thought as somehow outside of history. It is not just MMA, but sports in general which profits from a popular conception of being "natural" or "original" to human culture, when in reality these are highly constructed and rationalized cultural forms. Venerable sports historian Allen Guttmann explains: "sports as we know them today are not the natural, universal, transhistorical physical activity forms they are commonly thought to be, played in roughly the same way by all peoples in all periods of human history" (2004, p.33). Despite this assertion, Guttmann still finds recourse to notions of unrecoverable origins being accessed or satisfied, especially with regard to football and combat sports. While the structure and organization of football has all the hallmarks of modern society—rules, regulations, records—Guttman asserts that "the emotional function of the game may be primitive, even atavistic" (p. 125). Football and rugby, for example, "afford an outlet to the primitive desire to bang into people" (Guttmann, p. 130). These points are significant in understanding how a creative marketing imaginary has tapped into narratives of sports' emotional function as linked to something "natural, universal, transhistorical," a "primitive" contents, if you will. This tracing of connections between the "primitive" and "natural" in modern sport is often tentative and open ended in scholarly literature.

Literature provides additional evidence of the naturalizing master narratives accompanying many of our modern sporting traditions. Joyce Carol Oates, for example, engages in a not uncommon form of romantic primitivism in her essays, *On Boxing* (1987). Oates goes to some lengths to inoculate the reader against her eventual primitivism, explaining that "because boxing is a story without words, this doesn't mean it has no text or language, that it is somehow 'brute,' 'primitive,' 'inarticulate'" (1987, p. 11). She is duly respectful of the boxer's art as accumulated cultural knowledge, and not "his merely human and animal impulses," yet the categorical conflation of human and animal here is in itself revealing (Oates, 1987, p. 15). Boxing "inhabits a sacred space predating civilization," rooted in "man's greatest passion" for war (Oates, 1987, p. 21–33). Furthermore, of the boxing spectator, Oates states that "the instinct to watch others fight and kill is evidently inborn" (1987, p.42). But perhaps Oates' most conspicuous activation of a naturalizing primitivism is when she suggests that, in our status as a wealthy, advanced civil society, boxing's existence is best sourced to the desire "not merely to mimic, magically, but to *be* brute, primitive, instinctive, and therefore innocent" (1987, p. 43–44).

In perhaps the best analysis of the development of MMA, Van Bottenburg and Heilbron (2006) explain the emergence of MMA as a process of "de-sportizing" traditional martial arts (Karate, Tae Kwon Do, Ju-jitsu, Wrestling, Boxing, etc.) by eliminating their accompanying rules, regulations, and organizing bodies for promoters to produce a more reality-based fighting spectacle. After promoters realized the limited market for such an extreme, fringe spectacle of violence, the sport—spearheaded by the UFC promotion—then underwent a re-sportization, adopting much stricter rules and regulations in order to gain wider market acceptance. Van Bottenburg and Heilbron explain the effects of de-sportizing as "as reduction of rules and regulations in pursuit of greater authenticity, blurring the boundary between martial arts and real fighting" (2006, p. 269). And indeed, the UFC presently markets itself under the tagline, "as real as it gets." Borrowing the concept of "sportization" from Norbert Elias (1971), Van Bottenburg and Heilbron adopt the argument that the history of sportization is linked to the "civilizing process." The civilizing process results in the "pacification of everyday life," where physical violence is permitted only in socially sanctioned events (2006, p. 263). In order for the civilizing process to succeed, physical violence had to be rendered socially, even morally, reprehensible, with the exception of certain socially sanctioned forms of combat. If the link between the sportization of physical combat and the civilizing process is accepted, it follows that MMA represents a de-sportizing, and thus on some level a regression from the social values of civil society.

Prior to emergence of MMA, sentiments among traditional martial arts practitioners and instructors revealed a common belief that excessive rules and regulations limited their respective combat forms. There was the sense that "overly tight regulation forced fighting styles too far away from their origins: as exercises for real fighting. A street fight did not stop if someone scored a point or a particular throw was used..." (Van Bottenburg & Heilbron, 2006, p. 267).

UFC 1 was held on 21 November 1993. Brazilian Royce Gracie used highly refined jujitsu skills to force several bigger, stronger men into submission, and won the tournament. At the outset MMA was, as Greg Downey puts it, "more technically sophisticated than instinctually savage" (2007, p. 202). Of course, given the absence of other such "open" or "free" fighting spectacles, it is plain to see how the UFC invited such classifications as savage, instinctual, barbaric, and primitive.

The visual spectacle of an MMA fight is integral to encouraging a myth of the "real" or "natural." Unlike traditional martial arts, like Karate, for example, where competitors wear a gi and are more fully clothed, MMA fighters wear only shorts (or spandex) and small gloves. And there is a greater level of physical intimacy than found in boxing, football, or traditional martial arts because of the grappling or "ground game" dimension. MMA fighters wear smaller gloves than in boxing or kickboxing, with open fingers to enable grappling techniques, and in this way MMA gives the impression of eliminating artifice as obstructions to "real" or "natural" fighting. Put another way, MMA is less mediated by gear or other introduced technical apparatuses that mitigate the potential variety of ways in which combatants can fight each other. The cage, for example, is not in itself more "natural" than, say, the ring in boxing, but is conceived to allow for a wider range of fighting styles and techniques (Downey, 2007).

The image of minimally clothed or gear-protected (i.e. armored) men punching, kicking, and grappling on the ground to the point of either submission or lack of consciousness surely suggests pre-modern, even pre-technological man's struggles for survival in a harsh, "survival of the fittest" environment. Yet as Greg Downey argues, drawing from the seminal work of Marcel Mauss (1973), one should not "assume that the (nearly) naked human body is not already a technological artifact, shaped by cultural training techniques and subject to social dynamics" (2007, p. 203). Downey's argument is that MMA has worked to create "the closest approximation of 'real' fighting permitted under the law," but is by no means "real" or "natural" in the absolute sense (2007, p. 206). For Downey, the important thing to appreciate is how training and technique, as it corresponds to the form MMA has taken (what he refers to as its "technological minimalism"), result in a certain kind of skilled, and new form of socialized body (2007, p. 211). That the categories of the "natural" and the "real" are "suspicious" to Downey—and yet necessarily anchor such an analysis to begin with, as the operative discursive construct offered up by the popular commercial imaginary—reinforces the necessity of critical inquiries into how these constructs manifest themselves in sports' symbolic visual culture (Downey, 2007).

"Sporting" Nature(s)

Julia Corbett (2006) poses the question, "When do wildlife make the news?" They do, she submits, "either when someone is making claims about them, or when the boundaries (symbolic or real) between humans and wild animals overlap" (p. 204). The advertisements and t-shirts analyzed in this section are just such cases of blurred symbolic boundaries between humans and animals (or culture vs. nature), yet ones which did not make the news. Tapout's television commercial "Eye of the Storm" (2008) capitalizes on a popular socio-psychological imaginary of the male inner animal, the eye configured as window to the immeasurable, unknowable potential prowess lurking within. In visual terms, the commercial positions the viewer in a first-person perspective across the cage from a fighter, bobbing and weaving in slow motion towards the camera, fists raised in attack readiness (See Figure 2). As we converge with the fighter, moving gradually from long-shot to extreme close-up, the view closes in rapidly onto a single eye. From here the framing is static, but the human eye actively transmogrifies through a sequence of four distinct types of animal eyes, and then with a satellite view of a swirling "eye" of a hurricane in perfect graphic match over the eye. The final image cuts back out again to medium close-up in order to show the Tapout brand MMA gloves.

In the Tapout commercial, human physiognomy is not attached to an animal (anthropomorphically), rather animal qualities are attached to a human (theriomorphically). The commercial represents an explicit theriomorphis, a wilding, animalizing,—or re-naturalizing—impulse that lies behind so many of the brand's t-shirt graphics and their symbology. To seamlessly superimpose photo-realistic images of animal eyes on a human face blurs the human–animal distinction, engaging a far-reaching ideological discourse of humanity vs. nature. These images have the function not only of assigning the natural world with certain meanings, but in reflecting entrenched cultural beliefs of our own relationship to nature. As suggested earlier, animals in advertisements tend to signify characteristics of physical prowess, such as power, speed, strength, but also such traits as bravery, spirit, and loyalty. Animals function here as "species representatives, not individual animals," and viewers require no special knowledge or expertise to apprehend the cultural values assigned to them (Corbett, 2006, pp. 165–207).

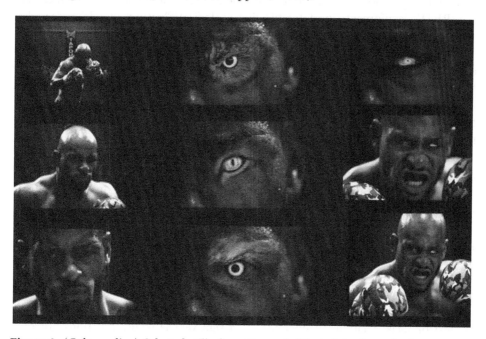

Figure 2. (Color online) Selected stills from Tapout's "Eye of the Storm" advertisement (author-produced montage).

Since the figure in this commercial is a black man, we cannot avoid acknowledging the specter of a long and dubious historical pattern of the white Western male's reliance on racial and cultural othering with which to (re)define their masculinity against. Historically, discourses invoking a lost primitive prehistory as a means of critiquing a supposed condition of over-civilization have been complicit in supporting white supremacy and maintaining hegemonic masculinity (Bederman, 1995). Yet the recent history of primitivism(s) has arguably been untethered from historically racist, colonizing ideologies. The "primitive," for example, was adopted by the 1960's Left in anti-technology protest, as inspiration for communal life, and in an attempt to deepen human–ecological engagement and raising environmental awareness. The postmodern discourse of the "primitive," and by extension "nature," is largely detached from grounded (geographic) cultural populations "out there," but is now instead, as Torgovnick aptly puts it, a "general marketable thing—a grab bag primitive," or free-floating signifier (1990, p.37). In the case of the Tapout commercial, the "primitive" is put to use in suggesting the male animal and inner (read socio-psychological) wildness as something out there to be tapped into. That wildness, aggression, and violence are part of our "nature," are "natural," yet also mysterious, unknowable, and in the realm of metaphysics. It is not surprising then, that in discussions of fighters, and athletes in general, there is always talk of one's "heart" or "spirit." "Nature" stands in for the romantically unquantifiable, or imaginary, dimension of human (athletic) performance. When so much of modern sport has become quantification through excessive statistics, historicizing (i.e. temporalizing) canons of great athletes/moments/events, and high-tech enhancement (i.e. P.E.D.'s, equipment, spectator viewing ecologies), "nature" stands in for what cannot be measured or quantified.

If you have watched events from The Ultimate Fighting Championship (UFC) or other smaller MMA promotions, such as Strikeforce, Pride, or Bellator, then you may have noticed the phenomenon of the MMA t-shirt worn by fighters and fans. The UFC now attracts large corporate sponsors, such as Bud Light, Harley Davidson, and Burger King, to name a few of the more ubiquitous sponsors. Beyond these, a host of brands selling gear and fan apparel have emerged, most notably: Tap Out, Bad Boy, Affliction, Venum, Dethrone, Xtreme Couture, Hayabusa, and Form, to name a few. All of these brands produce MMA imagery in the form of graphic design and advertisements, and they are displayed perhaps most prominently on t-shirts during fighters' ritualized pre-fight walk to the cage, in post-fight interviews, press conferences, weigh-ins, and related media. The MMA t-shirt—often produced as part of a sponsorship deal for fighters, but available for purchase by anyone—is the most common visual marker of MMA fan identification. The choice to wear, or "sport" fragmented "nature" is part of the same symbolic imaginary expressed in the Tapout commercial, but qualitatively different in being an embodied (i.e. worn) tableaux. T-shirts operates in material culture where its images work within the context of social relations; it operates as "a sign vehicle whose functions not only express selves, but the social and political fields in which it exists" (Cullum-Swan & Manning, 1994, p. 417).

In wearing an MMA t-shirt with thorn, wings, fangs, or animal eyes, one is, in a sense, self-primitivising, and perhaps subscribing to some common-sense notion that we are merely on a continuum with other animals in nature, and not categorically separate. A "naturalist" approach views the distinction between human and animal, or human and nature, as "matters of degree," where a "culturalist" position sees the difference as absolute, rooted in the symbolic capacity of humans (Soper, 1995, p. 81). Quite significantly, this calls to mind the crucial distinction between, on the one hand, invoking "nature" as a metaphysical (or spiritual) concept for thinking the "human." Or, on the other hand, as a realist concept referring to the (actual) natural world, or as Soper defines it, "the structures, processes, and causal powers that are constantly operative within the physical world" (Soper, 2000, p.125). The danger in exploiting metaphysical versus realist discourses of nature is in the potential to promote unsustainable human–ecological relationships, such as through the violence and aggression of combat / war.

In the MMA commodity spectacle t-shirts and skin are emblazoned with fragments of predatory animals and other "natural" iconographies, which might be productively understood as visual–rhetorical actions employed by companies aiming to cash in on certain consumer desires, anxieties, and dissatisfactions (Corbett, 2006). The phenomenon of the MMA t-shirt is also remarkable because of the sheer quantity of designs produced, and should be considered as an analogue to the myriad tattoos "worn" by fighters. The t-shirts are often graphically similar to fighters' tattoos, and with fighters more commonly tattooed than not, the t-shirt's visual iconography affords the fan greater symbolic vicinity to fighters' specific version of corporeality or embodiment. The MMA fan is thus enabled to ascribe or overlay themselves with a similar set of naturalizing signs and symbols as the fighter so often inscribes directly into their flesh. Similar to the manner in which an SUV offers a "vehicle" to reconnect with the wild from a position of relative safety, the t-shirt offers the MMA fan access to a vicarious corporeal re-wilding through the safety of the t-shirt as kind of second skin. That is, a t-shirt is easier to take off than a tattoo.

Yet the MMA t-shirt affords a different use and function for nature imagery than the more common, "nature as backdrop" mode commonly used in advertisements. One way to understand this difference is in what Panofsky (1982) understands as "open" versus "disguised" symbolism (as cited in Van Leeuwen, 2001). Open symbolism occurs when the visual motif is not represented naturalistically, as in the case of an MMA t-shirt relying on an expressive, graphic amalgam of motifs drawing direct visual analogies between the wearer and the "natural" world. Disguised symbolism, on the other hand, is when a visual motif is represented naturalistically, following laws of realism or verisimilitude. A typical "nature as backdrop" ad, such as the SUV or pharmaceutical advertisement, elides the symbolic role of "nature" or environment in its narrative by appearing incidental or merely contextual, and is thus more likely to be disguised to the uncritical viewer. A brand like Tapout, while also selling utilitarian MMA gear, specializes in selling t-shirts whose primary use is arguably to ascribe to the wearer through their graphic elements an ideological and lifestyle value connoting both physical aggression and defensiveness. As a part of the larger capitalistic spectacle of constructed/commodified "nature" / "the natural", MMA's construction of it is quite distinct from the predominant nature as backdrop format. While the SUV promises access to the wild/wilderness, it doesn't promise access to an inner wildness, animal, or romanticized origin state(s) associated with the masculine primitive. While MMA's greater availability to a naturalizing visual discourse is now evident, two recent Nike commercials extend the argument to commerce and sport in general.

The commercials—"Alter Ego" and "For Warriors Only"—made for Nike's line of "Pro Combat Apparel," function symbolically to problematize the merger of artificial tools and technologies with the natural technical means of the body. These commercials present a highly ambivalent view of the convergence of high-tech gear (the synthetic) with the body's "natural" (technical) means. This is accomplished by visually constructing the elite athlete to embody a version of "primitive alterity" or otherness through the device of blurring the synthetic and the natural.

In Nike's "Alter Ego" commercial, Minnesota Viking running back Adrian Peterson's skin is monstrously inscribed (through digital overlay) with the pattern of the Pro Combat apparel worn under his uniform (See Figure 3). The image is a striking digital fusion of the artificial and the "natural," or the technological and the human, and the message is just that—technology so similar to nature that the boundaries are blurred. If we did not first see the "deflex" pattern of the Pro Combat protective padding on Peterson, the mirrored pattern on his skin would otherwise register as a kind of reptilian armor. The image of Peterson registers as a human–

Figure 3. Selected stills from Nike's "Alter Ego" advertisement (author-produced montage).

nonhuman hybrid species. Football as a visual spectacle, much like MMA (though for different reasons), lends itself to primitivism's organizing conceptual tension of people looking back to origins to understand their present and potential future. Guttmann affirms this, explaining that football "announces the continuity between contemporary man and his most ancient ancestors;" and in a virtual echo of the commercial's images, football's "elaborate gear seems to emphasize both the primitive and the futuristic" (1978 p.125). It is not surprising then that the commercial enacts a tone of otherworldliness typical of the science fiction genre.

The commercial's mise-en-scene has a deep, inky, gothic tone with snow falling on the field of play that, as rendered in high contrast black and white, might as well be ashes from the apocalypse. Elite athletes glorified here as subjects on the margins of humanity, pushing the limits of human potential in primal, physical ways. This encompasses a lot things, but here especially the fetishism of technology, the breathing of life into "dead" technology so it becomes more than its mere instrumentality or functional logic (high-tech underwear in this instance). Always in tension in this scenario is the dystopian view of nature's technology as irrational, dark, monstrous, Frankensteinian, animal and libidinal, versus the utopian views of technology as rational, light, functional, contained and controllable, or comfortably "dead" (Rutsky, 1999).

The title of the commercial is more revealing, indeed appropriate, than one may guess. "Alter" is the Latin root of alterity, or "otherness," which implicates the "primitive" as the category par excellence for constructing cultural otherness. "Primitive alterity," Kurasawa explains, "has been pivotal in the process of Western modernity's constitution," and its primary role in the "symbolic domination" of others (Kurasawa, 2002, p.3). The "Ego" in "Alter Ego" also implicates an internalized/inner "primitive," understood as the shifting of primitivism's focus from the anthropological subject, out there in space and time, to the socio-psychological primitivism associated with modern alienation, and a desire to reclaim one's buried inner "nature" (White, 1985). Yet "Alter Ego" can also be seen to invoke the racial, colonialist primitivism that some have argued is now dead (Di Leonardo, 1998; Torgovnick, 1990). The commercial employs a somewhat disguised symbolism of "nature" as primitivity through an image of animalized skin on a black, racialized body. But the image and narrative it is contained within functions as a glorifying and aggrandizing masculine primitive, implicating a wider twentieth century "return of wildness," in this case manifested as sports advertisings' response to a consumer desire to engage in a "male animal ideology" (Bordo, 2000; Kurasawa, 2002). Yet at the same time as "Alter Ego" is glorifying the masculine primitive rooted in physical prowess and domination, it also evokes a dystopian questioning of the dangers of unfettered technological improvements of "natural" human performance. We see this ambivalence reflected in another Nike commercial for Pro Combat Apparel, "For Warriors Only."

"For Warrior's Only" articulates more explicit or open symbolic analogies between "natural" and "human," animal and athlete, than "Alter Ego," but reflects a similar ideological imaginary (See Figure 4). Where "Alter Ego" uses sophisticated digital compositing techniques to realistically render human skin into the form of a Nike product, "Warriors Only" uses rapid cross-cutting between images of nature and athletes to generate associative or suggestive meanings about technology's place in enhancing our natural physical endowments.

The commercial begins with a rapid-fire cascade of still images that will later constitute the climax of the commercial. The images flash at such a fast rate as to only allow the most minimal visual registration. These include masks and headwear functioning as a cultural mimesis of nature; the masks mimic a scorpion, a Venus fly trap, and a gazelle. Next we witness a slower but still rapid sequence through a series of professional athletes putting on Nike pro combat apparel in a stark, non-descript concrete industrial corridor. Following this literal "gearing up," a sequence of rapid cuts shows the athletes enacting the physical forms and motions typical of their sports, baseball and football. Next images of the athletes enacting their kinesthetic forms are intercut with images of the athletes with masks, without them, and then with actual images of a scorpion, Venus fly trap, and a gazelle. The sequence also includes one athlete wearing a head piece made of barbed wire, associated with containment, defensiveness, and danger, which does appear to be given any juxtaposition to a nature image unlike the other masks. The barbed wire does, however, like the other nature motifs, reference thorns, fitting into a category of "nature as threat" (Hansen, 2010).

Figure 4. (Color online) Selected stills from Nike's "For Warriors Only" advertisement (author-produced montage).

The visual analogy made by the athletes wearing masks resembling an actual referent in the natural world can be understood as the athletes acquiring, or accessing, a similar order of "natural" physical endowment as the wild species being depicted. That they are viewed putting on their Nike apparel before acquiring an order of prowess equal to the natural world suggests the synthetic or cultural mediator (i.e. the apparel) as the source of enhanced powers. In other words, Nike is fetishizing its products through direct associations with the forces of "nature." Again, this case stands as a quite open visual symbolic analogy of human to nature, where "nature as backdrop" ads functions in a more disguised or implicit manner. This tendency for deploying more open "nature" symbolism in the social arena of sports is significant, indicating that the marketing imaginary for sports constructs "nature" quite differently than other commodity forms. That is, the sports marketing imaginary on display here employs nature imagery in ways that express a more vexed or uneasy questioning of our human status, physical potentials, and perhaps even a sense of alienation from "true" or authentic selves through the aspirational technologies of sport.

Conclusion: Commodifying Nature/Metaphysical "Nature"

One of the central questions driving this analysis has been: what does the coupling of nature imagery with modern sport—our predominant cultural sites of socially sanctioned aggression—accomplish symbolically and rhetorically? And, using GCS' critical imperative of examining representational "nature" for impacts upon material (i.e. actual) nature as a guide, what human values do nature/naturalizing images in sports advertisements (taking the MMA t-shirt as, in effect, also an advertisement) promote, and to what potential ends? My contention is that this marketing imaginary is in alignment with popular or common sense notions of human aggression and violence as natural, and also that it operates in response to consumer desires to access or somehow consummate this knowledge. It is less a question of whether aggression is indeed innate, but rather to more fully appreciate the symbolic or discursive regimes the marketing imaginary is operating through. To be naturalized means, as Fabian explains, "to be separated from (historic) events meaningful to mankind" (1983, p.13). Thus to posit the human body, its power and prowess, within the timeless domain of "nature," as these visual discourses imply, is to naturalize it. Grace Jantzen points out that, "in the discourses of modernity, aggression has been taken as 'natural,' an innate feature of what it is to be human" (2002, p. 5). Countering this naturalization, Jantzen argues that (drawing on Bourdieu) violence has "colonized our habitus," and through prevailing Western "master narratives", be they theological (original sin), psychological (death drive), biological (testosterone), or political (competition, resources), they "render it theoretically inevitable and practically repeated" (pp. 5–8). I suggest here that the naturalizing imagery in MMA and the wider sports imaginary is also acting within such master narratives, reinforcing a habitus of violence.

MMA, unlike other popular mainstream sports, appeals through the (relative) appearance of minimized restrictions, rules and regulations typically associated with modern sporting institutions. It appeals on a "primal level," as a spectacle of survival, "instinct," and basic aggression (despite the highly technical martial arts used by athletes) intensified through its comparative "technological minimalism" (Downey, 2007). As a result, its marketing imaginary has adopted a naturalizing discourse in the form of the masculine primitive, with nature as primitivity. As UFC president Dana White has said on many occasions, "it's in our DNA." White states:

> Fighting—I don't care what color you are, or what language you speak, or what country you live it, we're all human beings and fighting's in our DNA. We get it and we like it. Before a guy ever hit a ball with a bat, or a ball went through a hoop, there were two guys on this planet and somebody threw a punch, and anybody who was standing around was watching it. It translates across all different barriers, and this will be the biggest sport in the world, I guarantee it. (as cited in Martin, 2010)

White's sentiment reflects a western master narrative of the naturalness of human violence, a biological determinism that is by no means clear-cut. In fact, biological evidence alone for male aggression is by no means conclusive, and studies suggest that prevailing cultural definitions and narratives of ideal masculinity are more closely associated with male aggression (see Kimmel, 2000, for a detailed synthesis of this research). Jantzen (2002) argues that most of the dominant western narratives (or "master discourses"), and as I attempted to tease out in the literature of sport, reductively assume rather than prove that human violence is intrinsic, instinctive, or simply human nature. In the process these narratives continually naturalize violence, and thus society is held in the "grip of a dominant symbolic system without bringing it to critical scrutiny" (Jantzen, p. 6).

These master narratives were also implicit in Marx's positing alienation from our "species being," and our "field of significance" (Corbett, 2006; Marx, 1972). With businesses like the UFC finding, as they have, a lucrative marketing synergy between cage fighting and The U.S. Marines, their success in romantically naturalizing masculine violence—and the convenient use of social Darwinistic rhetoric in codifying the sport—is certainly paying off. But this marketing imaginary works through exploiting a spiritual, metaphysical conceptualization of the naturalness of human-on-human violence, and arguably not a realistic or sustainable one. "Nature" and what is "natural" can be variously drawn and configured, but our "species being" or "field of significance" narrated in these images need not compose the "human" in terms of violence and aggression, though they are undoubtedly savvy business propositions.

Television Advertisements Analyzed

Company Title URL

Nike "Alter Ego" http://www.youtube.com/watch?v=EbnQL9mvFQQ
Nike "For Warriors Only" http://www.youtube.com/watch?v=EGOSZa0fg8Y
Tapout "Eye of the Storm" http://www.youtube.com/watch?v=KxGUaZTRGa4

References

Bederman, G. (1995). *Manliness & civilization: A cultural history of gender and race in the United States, 1880–1917.* Chicago, IL: University of Chicago Press.

Berger, J. (2009). *Why look at animals?* London: Penguin.

Bordo, S. (2000). *The male body: A new look at men in public and in private.* New York: Farrar, Straus and Giroux.

Corbett, J. B. (2006). *Communicating nature: How we create and understand environmental messages.* Washington, DC: Island Press.

Cox, R. (2010). *Environmental communication and the public sphere* (2nd ed). Thousand Oaks, CA: Sage.

Cullum-Swan, B., & Manning, P. K. (1994). What is a T-shirt? Codes, chronotypes and every-day objects. In S. H. Higgins (Ed.), *The socialness of things: Essays on the socio-semiotics of objects* (pp. 415–34). Berlin: Mouton-de Gruyter.

Diamond, S. (1974). *In search of the primitive: A critique of civilization.* New Brunswick, NJ: Transaction Books.

Di Leonardo, M. (1998). *Exotics at home: Anthropologies, others, American modernity.* Chicago, IL: University of Chicago Press.

Downey, G. (2007). Producing pain: Techniques and technologies of no-holds-barred fighting. *Social Studies of Science, 37*(2), 201–226. doi:10.1177/0306312706072174

Elias, N. (1971). The genesis of sport as a sociological problem. In E. Dunning (Ed.), *The sociology of sport: A selection of readings* (pp. 88–115). London: Frank Cass.

Fabian, J. (1983). *Time and the other: How anthropology makes its object.* New York: Columbia University Press.

Guttman, A. (2004). *From ritual to record: The nature of modern sports.* New York: Columbia University Press.

Hansen, A. (2010). *Environment, media and communication.* London: Routledge.

Hochman, J. (1998). *Green cultural studies: Nature in film, novel, and theory.* Moscow, ID: University of Idaho Press.

Jantzen, G. M. (2002). Roots of violence, seeds of peace. *The Conrad Grebel Review, 20*(2), 4–19.

Kimmel, M. S. (2000). *The gendered society.* New York: Oxford University Press.

Kurasawa, F. (2002). A requiem for the "primitive". *History of the Human Sciences, 15*(3), 1–24. doi:10.1177/0952695102015003165

Lasch, C. (1978). *The culture of narcissism: American life in an age of diminishing expectations.* New York: Norton.

Lovejoy, A. O., & Boas, G. (1965). *Primitivism and related ideas in antiquity: Contributions to the history of primitivism.* New York: Octagon Books.

Magdoff, J., & Barnett, S. (1989). Self-imaging and Animals in TV Ads. In R. J. Hoage (Ed.), *Perceptions of animals in American culture* (pp. 93–100). Smithsonian Inst Press.

Martin, D. (2010, February 18). Dana white: Fighting's in our DNA. *MMA Weekly.* Retrieved March 4, 2010, from http://www.mmaweekly.com/dana-white-fightings-in-our-dna-2

Marx, K. (1972). *The Marx–Engels reader* (Vol. 4). New York: Norton.

Mauss, M. (1973). Techniques of the body. *Economy and Society, 2*(1), 70–88. doi:10.1080/03085147300000003

Meister, M., & Japp, P. M. (2002). *Enviropop: Studies in environmental rhetoric and popular culture.* Westport, CT: Praeger.

Oates, J. C. (1987). *On boxing.* Garden City, NY: Dolphin/Doubleday.

Panofsky, E. (1982). *Meaning in the visual arts.* Chicago, IL: University of Chicago Press.

Pollay, R. W. (1986). The distorted mirror: Reflections on the unintended consequences of advertising. *Journal of Marketing, 50*(2), 18–36. doi:10.2307/1251597

Rotundo, A. (1993). *American manhood: Transformations in masculinity from the Revolution to the modern era.* New York: Basic.

Rutsky, R. L. (1999). *High techne: Art and technology from the machine aesthetic to the posthuman.* Minneapolis, MN: University of Minnesota Press.

Soper, K. (1995). *What is nature? Culture, politics, and the non-human.* Oxford: Blackwell.

Soper, K. (2000). The idea of nature? In L. Coupe (Ed.), *The green studies reader: From romanticism to Ecocriticism* (pp. 139–143). London: Routledge.

Torgovnick, M. (1990). *Gone primitive: Savage intellects, modern lives.* Chicago, IL: University of Chicago Press.

Torgovnick, M. (1997). *Primitive passions: Men, women, and the quest for ecstasy.* New York: Alfred A. Knopf.

Van Bottenburg, M., & Heilbron, J. (2006). De-sportization of fighting contests. *International Review for the Sociology of Sport, 41,* 3–4. doi:10.1177/1012690207078043

Van Leeuwen, T. (2001). Semiotics and iconography. In T. Van Leeuwen & C. Jewitt (Eds.), *Handbook of visual analysis* (pp. 92–118). London, IL: Sage.

White, H. V. (1985). *Tropics of discourse: Essays in cultural criticism.* Baltimore, MD: Johns Hopkins University Press.

Mobilizing Artists: *Green Patriot Posters*, Visual Metaphors, and Climate Change Activism

Brian Cozen

This essay examines the Canary Project's Green Patriot Posters *campaign as activist art that collectively comments on the cultural coherence of our current relations to the environment, particularly in terms of global warming, sustainability, and the concept of linear economic growth. Aspiring to bring together artists under the eco-activist umbrella, the Canary Project relies on an old WWII-inspired frame with a narrow premise of that period's conservation efforts. Within this framework, a range of visual designs question, subvert, and promote continued economic growth and an ontology that "more" equals "better." An analysis of the up–down orientational metaphors underscores a typology of these valuations and reveals one way to assess implications of such artistic efforts. That is, artistic expressions adapt and play with the contingent nature of metaphors, offering elaborations, extensions, and alternatives on basic structural elements and, hence, remark on how we orient ourselves and productively imagine being of/in the world anew.*

In the low-lying Pacific island, Kiribati, Lewis (2010) visually illustrates the human faces imminently impacted by rising tides. One can similarly see the impact of a shifting climate with Hurricane Katrina. Warmer water increased the intensity of the hurricane, and the water rose above and drowned the bodies and homes that were left below. In each case, more (water, temperature) devastates. Responses to such trends relate to the valuations of economic activity that perpetuates such catastrophes: we must maintain/increase consumption through increasing our efforts at alternatives,

---

Brian Cozen is a doctoral student in the Department of Communication at the University of Utah.

says the reformist; we must question what increases with increased production, says the skeptic; and we must reorient our practices around a more cyclic model that sustains human health in relation to an ecological system in which we are interlinked, says the radical. That is, environmental activist responses to the threat of climate change are numerous, and range from reformist to radical. We can see such responses in our visualizations. Hansen (2011) suggests two underserved avenues for research on visualizing the environment: how visuals are "made to mean" within multimodal, narrative contexts (p. 17), and "how visualization and the construction of visual meanings serve to bolster and privilege particular ideological views and perspectives on climate change over others" (p. 18). In the tension between global capitalism and environmental activism, ideological orientations to progress and the economic growth model are of particular importance.

Here I turn to activist art that both perpetuates and contests an economic growth orientation and an ontology that "more" equals "better." This essay examines the Canary Project's *Green Patriot Posters* campaign in order to empirically analyze how visual meanings are constructed around such ideological views and perspectives. The Canary Project, "an arts collaborative that addresses climate change and the need for a new economy" (Canary Project, "Public notice"), is a joint effort between photographer Susannah Sayler and her partner, Ed Morris. Inspired by Kolbert's (2006) *Field Notes from a Catastrophe*, the project began with Sayler's *A History of the Future*, photographic images of climate change ruin – such as Plaquemines Parish, LA, 3 months after Hurricane Katrina[1] – and potential solutions – such as the Dutch safeguards from their low-lying topography – in an attempt to visualize climate change throughout the world. The Canary Project has since offered a platform for varied artistic, eco-activist visualizations, involving artists, designers, writers, educators, and scientists in multiple research-intensive works to help visualize global warming and sustainability issues. Current efforts include global interviews of people's reactions to the impact of climate change on their particular landscapes, as well as a public installation of the potential future water-line cutting through New York City. The general pattern of the Canary Project's role is one of assistance: in terms of financing and resources, networking and dissemination ("Canary project").

Green Patriot Posters is their self-proclaimed effort to highlight the activism side of the acknowledged tension between dual impulses: artistic expression and activist goals (Morris & Sayler, 2008). Sayler has stated that, while the activist goals were always apparent, they now see the role of photography – originally seen as documenting "visual evidence" to get "rational people" to see and believe – as eliciting emotional responses. Still, Sayler contends, rational mobilizations are also necessary to inspire people to action; to this end, the photographer points to *Green Patriot Posters* as the project's "squarely advocacy" effort which "seeks to find…a strong…upbeat messaging, in design, to promote sustainability" (Boston Society of Architects, 2010). With the purpose to influence perceptions and mobilize solutions, artists were invited to design posters promoting sustainability, and many have been disseminated in such venues as a book project (edited by Morris and Dmitri Siegel), on busses (the original effort, by Michael Beirut, was nominated for the Cooper

Hewitt National Design Triennial; see Cotter, 2010), through magazines (Metropolis, 2010), and in gallery exhibits (including one exhibit from 2012 solely dedicated to the campaign: "Public Notice: A Green Patriot Poster Salon"). In creating designs for the campaign, artists are explicitly invited to take inspiration from the past, evidenced by the campaign's name: that is, from WWII posters mobilizing efforts on the home front. Believing that climate change activism "needs more positivity and urgency" (Morris, quoted in Thill, 2010), war posters, having mobilized masses in a common cause, are to provide that today, to "encourage all U.S. citizens to build a sustainable economy" (Green Patriot Posters, "About").

I argue that, through a visual analysis of primary, orientational metaphors, environmental activists' varied artistic responses, ranging from reformist to radical, are present in these activist posters on global warming and sustainability; and can be analyzed in terms of how artists visually characterize up–down spatial metaphors, or how the visual medium spatially represents what is up, what is down, and/or what is conceived as cyclic in nature. Artistic reorientations of the underlying, conceptual metaphors of capitalist values can highlight new ways of thinking and being in the world, highlighting activist art's political implications. Up–down markings are strong *orientational* indicators and hence fruitful avenues to consider such re*orientations*. Elkins (1998), for instance, comments on abstract painters' frustrations with the viewers' engrained sense of gravity; if these artists include any blue in the top of the canvas, a viewer will often associate the top portion with an upward orientation, and the blue with the sky. In the ways that still media artists placate to such readily connoted conventions as well as manipulate their semiotic associations, such up–down markings are beneficial means toward an economic analysis, in that what is up and what is down visually are metaphorically associated with such conceptual entailments. Below I outline up–down primary metaphors and their entailments, as well as how their visual articulations are adapted to alternative, activist values. With *Green Patriot Posters* in particular, the war inspiration frames the campaign in a way that may diminish certain artistic expressions toward alternative ways of being in the world.

In the subsequent critique, I consider how these posters are "made to mean" (Hansen, 2011) particularly through up–down sensorimotor metaphors, orienting a viewer's particular relations to ecological issues. First, I argue for visual analyses of up–down orientational metaphors and how it relates to articulation theory. The articulation of such orientational metaphors underscores the political impulses and implications of activist art. It is important to consider, as Charteris-Black (2004) contends, metaphor's pragmatic dimensions along with its semantic and cognitive dimensions "in terms of its ideological and rhetorical components" (p. 2), and in the next section I turn to the particulars of the campaign to consider orientational metaphors in praxis. That is, I consider how *Green Patriot Posters* also articulates environmental activism through its inspiration from WWII posters, and that linkage's attendant concerns. Third, I analyze up–down visualizations in three subsections: (1) posters that problematize an idea that more is better by inverting it; (2) those posters that do more than invert the binary, pointing to alternative cognitive maps by defying up–down visual alignments and orienting the design around cyclic and root imagery; and (3) those that do not problematize an idea that more is better, and its attendant economic implications. I argue that the two latter sections – the first discouraged and the second (perhaps unwittingly) encouraged – reflect entailments to the war frame, and so I conclude with implications in terms of the values and orientations encouraged via visual metaphors.

Metaphors and Articulation

I will focus on one case study to consider the political role of artistic expression via metaphoric, visual manipulation. I contend that one can analyze different orientations to the economic growth model through looking at up–down spatialization metaphors in such visual displays as the *Green Patriot Posters*. That is, how do the artists orient the visual design, and what can that say about different conceptions of environmental issues and progress? Lakoff and Johnson (1980) pioneered ways to consider metaphor as the conceptual basis for thought and possibilities for articulating new ways of experiencing the world. They argue that our bodies, oriented to the physical world, associate sensorimotor experiences (the source domain) to our subjective experiences (the target domain). The authors state how MORE IS UP correlates to our embodied experiences: as we see more contents added to a container, the level of contents goes up. Concepts that principally characterize a good life are also characterized metaphorically in terms of an up–down spatial orientation: "GOOD IS UP gives an UP orientation to general well-being, and this orientation is coherent with special cases like HAPPY IS UP, HEALTH IS UP, ALIVE IS UP, CONTROL IS UP" (p. 18). (They also discuss FUTURE IS UP, important later on.)

Subsequent work, including Lakoff and Johnson (1999), focused on delineating lists of experientially grounded primary metaphors as "universal" embodied experiences that leave us with "no choice in this process" (p. 58) of acquiring them, and how complex metaphors are made up of hundreds of these primary metaphors. In visual analysis of primary metaphors, Ortiz (2010) takes up these universalizing claims: how primary metaphors are universally learned because our corporeal experiences are universal, that we cannot avoid it, and that these *metaphors* "are inherent in human nature and are the result of the nature of our brain, our bodies, and the world that we inhabit" (p. 164). Yet others also look to the cultural manifestations of metaphors (Kövecses, 2004; Yu, 2008; also see Bowers, 2009), illustrating what Gibbs (2008) terms the "paradox of metaphor": between metaphor as contingent and as "rooted in pervasive patterns of bodily experience common to all people" (p. 5). Indeed, Lakoff and Johnson's (1980) initial sub-cultural redefinition of values and historical contingencies points to how groups and eras will favor one aspect of "up" over another, so that certain values will take precedence. Indeed, MORE IS BETTER is a cultural entailment to MORE IS UP and GOOD IS UP that perpetuates the progress myth of linear economic production.

It is through this contingent aspect of our metaphorical concepts that we can reimagine our embodiment in the world. For environmental activism, as opposed to focusing on those primary metaphors that are *universal* to humans, helpful in metaphor theory is how metaphors' contingent uses can rhetorically mediate new ways of being in the world. That is, artistic expressions that adapt and play with the contingent nature of metaphors, elaborating or complicating meaning through such embellishments (as in poetics; Lakoff & Turner, 1989), can help reorient how we see our embodiment in the world. I do not want to simply take primary metaphors as the universal building blocks through which complex concepts develop, but instead wish to claim that we can play with these as well, or foreground them in some ways and not in others. In any case, it is helpful to remain open-ended in how cognitive links might shift in relation to orientational metaphors, in an effort to think through how we can not only reorient how we think but also how we see our bodies oriented in the environment (see, for instance, Abram, 2010). Even if commonalities are widespread, articulating alternative orientations point toward (needed) change. On these terms, many social movements, especially (radical) environmental social movements, are not simply trying to get policy change toward, say, alternative energy, but are trying to redirect our orientation to our bodies and our worlds (each plural).

I contend that primary metaphors are a good basis for analyzing how, through the play of visual elements, green art attempts to reorient the viewer's relationship to environmental concerns specifically stemming from the economic growth model. DeLuca (1999) combines ideographic analysis and articulation theory to illustrate how environmental activists have challenged the articulation of the ideographs "nature" and "progress," prompting "changes in social reality through changes in meanings" (p. 35). In other words, DeLuca turns to ideographs to measure how key terms are used contingently in public discourse, but he also states that images, metaphors, and narratives are other elements the critic can analyze. I argue that imagistic metaphors, as conceptual links, and articulatory practices, as linking elements, are similar concepts, and the disarticulation and rearticulation of primary metaphors, like the ideograph, offer similar potential toward reconstituting reality: to imagine a physical basis that, from the beginning, is cultured and historically situated. The rhetoric of social movements that attempts to break and rearticulate these "metaphors," and hence relations to the world "we live by," are important to consider in an age of ecological crisis.

Articulating War Efforts

In addition to the articulation of visual metaphors, the war frame—at the fore in the *Green Patriot Posters*' title and inspiration—suggests a particular articulation of environmental issues. The campaign's website acts as an archive for marshaling a shared vision, inviting any and all artists to upload their content which is then approved by an editorial board and available for circulation, either by users or the editors themselves. To the latter, Siegel and Morris (2010) compiled 50 of these submitted images into a poster book, calling on its consumers to "let it enter the culture" (p. 11), reflecting Harold's (2004) analysis of the INFKT Truth campaign. With the latter – pullout advertisements in magazines – and the former – detachable and ready to hang posters compiled in a purchasable book – in both instances the campaigns ask the reader to destroy its pages and create their own public messages.

While the editors claim, "there is no one prevailing ethos, aesthetic, or message" (Siegel & Morris, 2010, p. 11), there is an *attempt* at a prevailing aesthetic (and hence, overarching message): a call to get inspired by WWII government posters. On the website as well as the preliminary pages of the book, the creators of *Green Patriot Posters* explicitly state how the campaign's title and inspiration comes from government posters during World War II that promoted conservation during the war effort. This aesthetic form entails a value orientation: "It is amazing to realize that we have already been through this. Successfully" (Green Patriot Posters, "About"). Presumably, through such efforts as the disseminated poster, we can mobilize the public to conserve, just as we did during WWII: "Use it up–Wear it out–Make it Do! Our labor and our goods are fighting" (Green Patriot Posters, "Inspiration"). Through this inspiration, many of the submitted posters appropriate both visual and discursive aesthetics from the WWII era; for instance, in one poster, a revolutionary-era patriot rides a bike, and the copy reads, "To arms! To arms! Climate change is coming!" ("To arms, to arms").

This ideological orientation rests on a narrow premise of the wartime efforts. These conservation efforts had nothing to do with decreased consumption but were meant for *redirected* consumption. That is, the public was called upon to conserve so that the war effort could consume *more*, requesting a temporary compromise from the public – and promising a golden age of consumption that would follow a successful war campaign (see Cohen, 2003, pp. 69–73). In sum, the war frame suggests calls for redirected consumption, not a reorientation around, say, a land ethic. In promoting a "muscular" green movement and "creating a new cornucopia of abundance... [like our parents, in order] to preserve our way of life" (Friedman, 2007/2010, p. 13), as Friedman's excerpt does in the preliminary pages of the book, the war frame suggests an orientation tied to the same model of accelerating our consumption anew.[2] The book's overarching themes (if not of all the posters contained within) wish to attack not the patterns of capital exploitation but merely its perceived excesses.

Examining anti-toxic movements' articulation of their cause through films with "sexy" celebrities (John Travolta in *A Civil Action*, Julia Roberts in *Erin Brockovich*), Pezzullo (2006) considers both the positive exposure as well as their negative potential to "eclipse" (p. 28) the issues and collective struggles. In the subsequent analysis, I consider how the framing of the *Green Patriot Posters* around World War II government posters, a rhetoric of strength, and the overall character of the visual spatialization of up–down metaphors can potentially "eclipse" the more radical potential of art activism. This radical potential was present in some of the website submissions that did not make it into the subsequent book, with the book more explicitly adopting the war frame. The analysis, therefore, builds a typology of visual categories articulated in the posters while considering – by way of linking environmentalism to concepts, "war" and "patriotism," heavily invested in economic growth – the potentially limited circulation of some categories.

Green Patriot Posters and Their Visualized Values

Below I divide my analysis of up/down spatial metaphors in *Green Patriot Posters* (in the book and the website) into three parts: (1) posters that utilize up–down imagery in order to subvert its cultural connotations, questioning its metaphorical entailments; (2) marginalized posters that favor a design that utilizes imagery of cycles or roots, offering alternative primary metaphors on which to build complex metaphors; and (3) those that rely on up–down imagery to perpetuate consumptive values, if in redirected forms. I should add a disclaimer that this framing of an analysis (looking to up, down, and cyclic/root visual metaphors) is not nearly inclusive of all of the posters available, nor even all-inclusive of the force of those represented; however, I contend that not only are these visual metaphors widespread, but they also communicate a core set of ideological commitments integral to environmental social movements: namely, the question of progress and growth, and competing orientations that question either manifestations (remedying its negative attributes) or the model itself (attacking its patterns).

Up is Down: MORE IS UP, But What's Up?

Art that problematizes consumption often rearticulates the MORE IS UP metaphor and questions the cultural coherency of MORE IS BETTER (or more production and consumption as good, with both metaphorically linked through up–down spatializations). Here we see the most readily apparent subversion to consumptive values in the *Green Patriot Posters*, and below I examine some of its manifestations.

In "Upside Down," a poster in the preliminary pages of the book but without its own page, the artist folds the MORE IS BETTER associations by utilizing one visual metaphor Ortiz (2010) identifies: the fusion of objects – here, smokestacks facing one way and trees facing the other – into one hybrid image. This image does not usurp the up–down spatial orientation but points out the conceptual entailments' inconsistencies with other current, cultural values. Against a yellow backdrop, we see three black, industrial smokestacks, and the word "upside" in the black smoke rising above. Embedded within the spaces between the smokestacks is the optical illusion of two tree trunks, positioned upside down. Instead of the smokestacks' base, these "trunks" extend into a continued image of two trees, as well as a cutout of a bird, with the word "down" written below – or above if you flip the poster around; as the website's caption encourages, "Let's turn it around." The value system the art piece espouses functions in a capitalist world where more industrial growth, and hence more pollution, is given primacy; and the piece calls out the incompatibility of more (pollution) positioned on top and good (nature) cast in the dichotomous, lower position. While critiquing the current framework, it maintains the nature—(industrial) culture dichotomy.

Other examples question our cultural detachment from consumption chains, associating upward movement with wasteful practices. "Keep Buying Shit" is an image of a landfill, and the sheer accumulation of waste – presumably continuing an upward trajectory – dominates the canvas. "Don't be Stuck Up," stylistically inspired by WWII posters, features three light switches all in the on – or up – position, with arrows pointing downward, curved to suggest the circular motion of a finger to turn them off. In a metaphorical system coherent with economic growth and progress, the bearer of industrial light – the turned-on light switch – would be "up," but the poster's playful design suggests that wasting energy implies being "stuck up," inverting UP IS GOOD. Instead, leaving them on (up) means more wasted electricity, metaphorically associated with someone "stuck up," or detached from the realities of our consumption, and potentially asking: Is industrial society stuck in the on-mode?

While these two examples have their own detachable poster, the book resigns "Mugs are Great" to a portion of the opening page. Here we have 35 disposable coffee cups stacked in a pyramid shape, juxtaposed next to a mug, and a detached hand holding a sign that points downward and states, "mugs are great." The artist uses asymmetric object alignment (Ortiz, 2010) to point out how mugs are less (lower in the image, and the sign pointing out this asymmetry) and therefore better. Again, the call is not to conserve consumption, at least in terms of coffee, but to consider what goes up, or what we get more of, when we accumulate waste within our common consumption practices. Furthermore, "What Will Your Children Drink," not featured in the book, problematizes the FUTURE IS UP metaphor by suggesting that we are leaving our children with more and more chemical pollutants. In the image, a filled container aligns "more" with verticality. The container includes detergent brands – Oxiclean, Drano, Tide, and so on – and the title suggests that the future for our children is becoming worse as we continue to pollute our waters.

Finally, a few examples offer visualizations of a major catastrophic forewarning of global warming: rising sea levels. In "Water's Rising," 21 years, each in 5-year intervals from 2000 to 2100, are written out in seven rows. While "2000" has a blue line below it, the line increases in size incrementally until, by "2100," the number is nearly completely flooded over, and even by "2080," none of the numbers are decipherable outside of enthymematically filling in the next sequence. While the verbal message below ("The water's rising. Rise up to stop it") combines war rhetoric, suggesting that we must fight MORE with BETTER, the visual leaks beyond this discursive containment by suggesting a need for more radical measures. THERE WILL BE MORE IN THE FUTURE (Lakoff & Johnson, 1980, p. 22) takes on new meaning. The image of increasing water levels gives precedence to the FUTURE IS UP, fighting against the values entailed in other conceptual metaphors, such as economic progress, that threaten that future.

In "S.O.S.," we see the same thing, and like "Water's Rising," and here more explicit discursively, we see the connection to the metaphorical concepts, THE FUTURE IS UP and THE FUTURE IS MORE. Here, the "future" is visualized by the actual word in block lettering. Against a cold, sharp-edged, gray backdrop, the blue lower portion takes up about half of the image and partially covers the lettering. If blue symbolizes water, then our future is drowning, and hence the emergency warning: "S.O.S." The image evokes two questions: what will save our future, and what moves up? Rising tides do not merely signal water going up but also the result of industrialism and consumption continuing up; therefore, what is progressing?

These examples show that, while not all were in the book, and while some from the book incorporate the war frame potentially at the expense of fixating its own visual complexity ("Water's Rising"), there are still instances of subverting consumption values by foregrounding waste as "more." However, aside from two posters illustrated below, most of the examples in the next section were not offered their own detachable page in the book, suggesting that while disarticulations of the up–down metaphor may be readily apparent in broader art activism, the framing around a "muscular" green movement may have largely restricted metaphors that structure a more cyclical, interconnected orientation between humans and our environments. The final section will consider how the inspirational war frame can align with the (problematic) conceptual domains tied to MORE IS BETTER.

Interconnected Imagery: Alternative Conceptions to the Up–Down Binary

Here I turn to some poster examples that take more of a cyclic approach in their visual metaphors and, I feel in turn, suggest a way to reconsider binary orientations to the environment reflected in up/down visualizations, instead promoting alternative metaphorical expressions. The *compost metaphor* offered by Donoghue and Fisher's (2008) cultural performance, one such metaphor that fits some of these examples, suggests an environmental activism that promotes growth through struggle, labor, and the cyclic vitality of life focused on socio-ecological sustainability (Larson, 2011). Some of these themes are implicit in the poster "Sustain," even if the appropriated aesthetic offers an up–down image. Specifically, the image appropriates a war poster in which a disembodied arm, raised and vertical, holds scraps in its clenched hand; here, the clenched fist holds a carrot against a green background. "Sustain" shows strength of character in terms of struggle. The whole image is green except for the carrot's edible root and outlined forms, suggesting a call to "green" ourselves (the upward, raised human fist) in sustaining life's foundations. The toiling hand of the worker metonymically stands in for the moral strength of humanity, and while it perpetuates an up–down image, the poster also suggests the life-sustenance practices, and the toil through struggle, of the compost metaphor.[3] Below I consider two trajectories that consider this metaphor in two additional ways: the interconnectedness of humanity within an ecological system, and the cyclic nature of that interconnectedness calling to question the cultural myth of sustained linear growth.

Many circular images abound in the submitted posters, though largely only the ones with bicycle imagery made the poster book. Part of the reason may be in the editors' assessment of the bicycle: "a nonthreatening, non-ideological image, unsanctimonious and almost childlike" (Siegel & Morris, 2010, p. 10). However, bicycles are hardly non-ideological, as suggested by the poster title, "Join the

Revolution." Bicycles are indeed a revolutionary shift in physical *movement*. As McGee (1980/2006) points out, the "movement" in social movement is a metaphor for physical motion (p. 118), but the bicycle as symbol elicits both actual movement and a revolution in consciousness.[4] Website posters further this cyclical image of revolving consciousness. For instance, "Recycle" positions the eponymous word upside-down, on top of a right-side-up, mirror image. With the tagline, "Flip and enjoy longer," the image invokes the metaphor of recycling, visually – re-cycle becomes reuse, the perpetual folding over of a cyclic image.

Below I examine five examples – two offered as detachable posters ("Lumberjack" and "Global Warming") – that illustrate how, through the metaphor of economic growth (up) as good, we are cutting down our own bodies' health. The first is a triptych: "Corn-Oil/Water/Global Warming." This poster series questions an alternative "solution" to the energy question: the utilization of corn as a source of energy for continued economic growth. In these images, corn stalks rise from within an oil barrel, a water tower, and a tree stump, respectively. The image suggests, and the online descriptions contain as much, how the effort to expand corn-production perpetuates the wasteful consumption of fossil fuels, water, and redirected land use. These images do not so much flip the up–down metaphor as much as demonstrate how consumption and production interlink. Avoiding a visual distinction between the top half and bottom half of the design, consumption remains embedded within practices. Unlike, say, the poster "Mugs are Great," which critiques the specific ways to consume coffee but does not, for instance, comment on the crop's relations of production, the triptych points to the interconnected environmental consequences of such production practices. In energy policy, the focus on upward growth further risks other resources upon which we are interdependent, especially water (Cozen, 2010).

Two images, the book's "Lumberjack" and the website's "Ecology," point to this dependency by offering a visual hybrid of our bodies and our land, questioning the instrumental ontology industrialism favors. In "Lumberjack," a Paul Bunyan figure, ax in hand, looks down and sees a tree growing out of his foot. Essentially, the Paul Bunyan mythos chops off the feet from under us. In the hybrid image of "Ecology," a human form fuses with a tree, particularly the body in tree-form. The arms are raised above the head with an ax in hands, with movement and tension in the body signaling motion and the inevitable swing downward, ready to strike. The image captures the point where one more swing will cut down one's own tree-body, split open like a tree being felled. First, in a literal sense, the cutting down of trees affects our own selves. More broadly, and reflected in the title, "Ecology," humans and human bodies are interrelated with the ecological system: we are *of* the vitality of the earth. To ignore this inherent connection, and to degrade the environment – to "use *up*" resources without ecological concerns – is to damage our own health, our own bodies rooted in, not on or above, the earth. In relation to "Lumberjack," "Ecology" is more visceral: the chopping has begun, indeed is nearly complete. As opposed to the former poster, it is not merely an outgrowth, an oddity in the way we chop down trees or a sudden recognition of this ecology that has grown onto our bodies and instrumental myths. Instead, the image reflects a long-standing way of living that chops down our *center*, our torso, the core of our bodies. It is in this sense that radical critiques get at the literal "roots" of environmental issues.[5]

When perceiving the poster "Global Warming" in its upright position, smoke emanates from industrial smokestacks; this is merely where the visual story, positioning how what goes up must come down, starts. The smoke begins to go up into the atmosphere, as our physical experiences point out. But the image is made complex through its cyclic imagery. The black smoke circles back down, and if we look at the blackened portion that can represent the ground of the industrial plant and turn our head, we see the torso and head of an upside down schematic male body, and from that angle the smoke stacks are the human's legs, his upper half covered by smoke. Up–down associations flip in numerous ways. The pollutants do not just go up into the air but enter a damaging cycle. With a focus on the upward expansion of production, and ignoring the waste byproducts, it is us who get flipped around, our heads buried in our own creation. The smoke has gone down to cover us and is circling back upward. The stick figure-like person gives it a gendered everyman feel, but also it reflects the male image on bathroom doors, opening up a potential reading of the waste from which we alienate ourselves, or a metaphorical link between industrial waste and human waste.

On the back of the poster, the artist explicitly likens this piece to a snake biting its own tail, an image illustrated in another poster, found in the preliminary pages of the book but without its own page. "Golden Snake" embeds a phrase within the image of a snake: "Nature provides a free lunch, but only if we control our appetites." This phrase, and with it the snake image, completes a full circle, the snake's head about ready to bite. The phrase alone may not be as obvious in its subversion of the up–down dichotomy as the hybrid image and its detailed description:

> The snake that bites its own tail is an ancient symbol for self-reflexivity or cyclicality. Plato described a self-eating, circular being as the first living thing in the universe[,] an immortal, perfectly constructed animal. The human species is now at the top of the food chain and dominating nature. Ignorance and greed now risk the balance of this self-sustaining [sic] cycle. (Green Patriot Posters, "Posters," p. 5)

To paraphrase, the poster critiques a focus on consumption as upward, or as furthering linear growth unabated. CONTROL IS UP is a myth. Our appetites – our consumption of resources – are not disconnected but tied to our own bodily health. Instead, control only functions within an ecological perspective, a cyclic metaphor that considers power relations in terms of ecosocial sustainability.

Up Over Down: MORE IS BETTER

This section considers how closely aligned are some current posters to the underlying, unstated motive of the WWII posters: namely redirected, accelerated consumption. Here we would expect to see visual spatialization metaphors that do not problematize how more is better, simply that we need to increase our efforts toward having more of alternative forms, specifically of energy. While such examples are somewhat minimal, a few are important to consider: Shepherd Fairey's "Power Up Windmill" (titled

"Clean Energy for America" on the website) as well as "Efforts Up!," "Alternative Living," and "Paint It White."

"Power Up Windmill" (the book's title) combines a cyclical image with up-down values, images, and verbal cues to undercut its core image's sustainable potential. From Shepherd Fairey, the same designer as the "Obama: Hope" posters, "Power Up" utilizes a similar coloring scheme, symbolically patriotic in its reds, whites (or tans), and blues. Fairey describes the picture and its patriotism succinctly on the back of the poster: "I created this windmill image as a patriotic symbol of the green-energy mission." Economic progress defines patriotism: windmills are patriotic by powering *up* America, not so much because they can represent a shift in orientation to how we live. The cyclic windmill image and the stars of the United States flag, superimposed in the clouds to give the propellers a sense of circular movement, contrast with the title, "Power Up America," at the end of the Z-shaped eye-movement pattern. The corner image includes the word "up" written in larger typeface, flanked by additional windmill images with blades pointing vertically, aimed directly at the word "power." Our patriotic, "upward" mission becomes redirected consumption and maintained growth: THE FUTURE IS (pointing) UP.

The windmill image alone can represent a cyclical relation to the environment, but the verbal and the corner image focus on its utility aspect. That is, the windmill is not so much a symbol of sustainability as a sustainable means toward powering our economy (conflated as America). In the windmill, politics and economics collapse, and the political mission becomes finding a way to "power up America": "Clean Energy for America" as powering our economy and, therefore, patriotic. While indeed a move in the right direction, the poster ignores how the way of life achieved through industrialized consumption is based on unsustainable resource extraction, or the energy "slaves" of fossil fuels (Shiva, 1990, p. 196). Therefore, it encourages competing discourses that claim we cannot redirect all our necessitated, naturalized consumption over to wind and, therefore, must continue with coal and oil.[6]

In "Efforts Up!" we see a suggestion that humans are the solution while detached from the problem. The poster, replete with red, white, and blue, divides in half: the top half, a red backdrop with white lettering, "Efforts Up!" and white arrows above pointing up; and the bottom half, a blue backdrop with white lettering, "Carbon Down!" and white arrows below pointing down. "Efforts" and "Carbon" are in block lettering (as are the arrows), while "up" and "down" are in cursive and stylized, adding emotional emphasis to the direction of the directives. Also, the top-heavy design features five arrows facing up and only three that face down. This asymmetry suggests we need to increase our efforts before we can decrease emissions, simultaneously intimating optimism (we can increase efforts) along with the weight of the issue (our efforts need to outpace the tide we are pushing against). While it promotes agency, the poster remains abstract and somewhat deflects the links between industrial output and carbon proliferation: as if we are only the solution. There appears nothing inherently wrong with this message, in that stepping up our efforts is an effective rhetorical plea, but the problem develops when efforts are conflated with stepping up (or expanding) production. When starting with industrialized consumption patterns as base assumptions, proposed solutions just try to outpace attendant "externalities" through increased technology.

WWII posters argued a need for our collective "efforts up!" in order to maintain the war engine, and similarly today we see a general discourse of needing more energy alternatives to maintain consumption and the economic engine. In "Alternative Living" as well as "Consider the Alternative," neither included in the book, both posters articulate corn in terms of power and a way to maintain the illusion that more is (always) better. The former, visually an upright piece of corn with a cord attached, suggests to "power up" an alternative energy source; the latter, similar visually, asks, "Empowering, isn't it?" However, these posters promote an alternative way to maintain our problems, reflecting the rhetorical practice of conceiving food in terms of energy and the detrimental environmental concerns, especially in terms of water, with that articulation (Cozen, 2010). Regarding a green movement as upward, muscular, economic movement, or as a continued focus on the upward expansion of utilizing new resources, masks the patterns of production that are environmentally damaging by largely painting over systemic issues, nowhere more literalized than in "Paint It White." Visually, the image of a cityscape rises from the hybrid image of a paintbrush, in a poster that promotes painting rooftops white in order to cool down cities. Here we see a solution through consumption, literally painting over deeper problems. While Pezzullo (2011) is right to suggest that all politics are impure in our consumerist world and that we must do something – for inaction is the only guarantee for further devastation – it seems beneficial to highlight artistic expressions (itself a consumptive act) that do not merely perpetuate increased consumerism, as we see here, but also, as previously considered, problematizes such relations.

Conclusion

This analysis has considered different ways in which eco-art activists visually represent orientations to a growth model, as well as the potential ideological interests supported and contained by the war frame of strength and a MORE IS BETTER mentality, as evidenced in *Green Patriot Posters*.

Specifically, I analyzed how artists utilize multiple up–down spatialization metaphors both consistent with the concept of progress and as rearticulations, reflecting the conflicting values (Lakoff & Johnson, 1980) articulated by this art. While such orientations suggest a different form of power, not necessarily a qualitatively better moral position (see the discussion of control in "Golden Snake," above), I am most interested in the productive imaginings of orientations anew. As Lakoff and Turner (1989) argue that the artistic value of poetry can largely be drawn from the artistic manipulation of metaphors, here we see that artistic, and activist, dynamism of visual design is apparent when analyzing orientational markings.

While this study does not refute the scope to which primary metaphors are learned, it at least questions assuming a universality that underscores some cognitive metaphor theory. Specifically to up–down metaphors, for those seeking a re-orientation to the world – a very different ontological grounding – the focus on *more*

(more is better, more conservation efforts, more strength: "more" visually affirmed in UP metaphors) might not be the best cultural marker to assume. Nerlich and Koteyko (2009) state that change occurs once people start to "think differently about a topic, to change old cognitive habits, and entrench new cognitive habits – to see things in a new light, in fact to create new ontologies. However," they add, repeating a common position, "new thinking has to be rooted in something already well-known and familiar to make the jump from old to new possible" (p. 219). Nonetheless, this stance should not be taken as a blind reliance on the old. Artistic expressions are elaborations, extensions, and *alternatives* on basic structural elements and, hence, how we orient ourselves to the world. That is, artistic expressions are not simply taking our primary metaphors and flipping the cultural coherence, but also, for instance, potentially thinking anew the reliance on the up/down dichotomy to begin with: either turning to other engagements with the world (as with cyclic imagery) or muddying the binary by showing the interconnections (the root imagery). If rhetoric can be understood as the mediating force between the semiotic and the material (Sloop, 2009), alternatives to living in the world and thinking of the world need to develop together in order to enact change. Bodies in industrial societies orient to the environment in particular ways, and metaphorical theory needs to not only consider how to think anew but also avoid assumptions that we cannot also be of/in the world anew.[7]

In summary, if environmental art is to examine the problematic relationship to economic growth – argued here as a key environmental issue (also see Larson, 2011) – it must address the ontological connection humans have with their environments, not seeing the world as external and unilaterally controlled but as interlinked. This essay has argued for an analysis of orientational, up–down visual metaphors as a way to assess symbolic representations of such orientations to economic growth, through the concrete particularities of one campaign.

To conclude on its additional implications, this campaign also promotes abstract imagery and a past war aesthetic. First, to abstraction, if green art is meant to join local and global problems (Hall, Bernacchi, Milstein, & Peterson, 2009), the editors are trying to promote images that can circulate regardless of (American) locality, encouraging visualizations of environmental issues that abstractly problematize (or reinscribe) the macro issues of capitalism and energy production, if not its specific manifestations. A focus on places also often function as micro-cases of macro inequalities, as the Kiribati example to start this essay can illustrate (Lewis, 2010) and as the Canary Project's other campaigns attempt. Further analysis would consider this dissemination strategy: of the abilities of macro, abstract imagery to create visualizations anew and to circulate in local contexts.

Second, the totalizing war frame threatens to lessen such local imaginings and diminish the aspect of environmental interconnectedness many seek to foreground. While perhaps helpful in mobilizing efforts around a shared, nostalgic longing for action, it is important to show where the WWII inspiration falters. This paper suggests three liabilities. First, as Ivie (1987) has argued, the efforts to mobilize against a common enemy may find that the enemy ups its efforts, and we perhaps see that in sutured logics of "all of the above strategies" as discussed with the biofuels posters. However, it is not the melodramatic framing that is of concern in this essay so much as the additional entailments to war and WWII specifically. So, second, delayed gratification is a questionable frame for sustainability. Nerlich and Koteyko (2009) point out that, since WWII, "carbon rationing" is an accepted cultural frame in the UK, but it is not so in the US; therefore, "Carbon *Rationing* Action Groups" (CRAGs) change to "carbon *reduction*" in the US (p. 216). Indeed, the US imaginary around WWII is one of delayed abundance, not staying within limits, and so the obstacles of the present are quite different if not oppositional to the past and its complicated cultural narratives. Last, the focus on American strength and Friedman's Green Economy likely devalues alternative conceptions that do not merely flip the binary but get at more radical issues. It is to these radical positions where we might wonder if up–down visualizations are appropriate.

Acknowledgments

The author would like to thank Danielle Endres for guidance throughout the project, as well as Laura Lindenfeld, the issue editors, and the anonymous reviewers for their helpful feedback.

Notes

[1] The two images widely disseminated are the remains of an uplifted truck and horse, tossed into trees by the hurricane.

[2] Rogers (2008) makes a similar point, noting Friedman's essay as exemplary of the masculinized rhetoric in current environmental discourses. The Green Patriot Posters book is an attempt at recontextualizing the essay, retracing its masculinized rhetoric for further dissemination. In this textual strand, the essay's function is to popularize muscular environmentalism. Also see Singer (2010) on Friedman's "Code Green" rhetoric as a neoliberal call to "green" American capitalism.

[3] Benford and Hunt (1992) note how images of clenched fists directly contradict central themes of the peace activists who reject such imagery. "Sustain" indeed acts as an example of clenched, tense, armed "strength," a ready-to-fight metaphor stemming from its WWII influence. But, in addition, the appropriation opens up an idea of "up" as rising from roots in the ground, and it is those roots, like the carrot, that are valuable, enriching, and life-sustaining. See Rogers (2008) for how the carrot can also represent effeminacy in discourses recovering hegemonic masculinity.

[4] See Jensen (forthcoming), for a critique of bicycle transport: foregrounding automobility, or autonomous motion, sublimating the valuation of public transport, and reinforcing social rationalities around mobility, producing (and controlling) mobile subjects. Also see Furness (2010) for the history of bicycles as paving the way for the automobile subject and, hence, positioned as both car culture's antagonist and its ontological and ideological origin.

[5] Madison (2010) describes this Marxist position well and reminds us that, while important to consider art's production – to see whether it addresses contemporary tensions – it is the relationship with an audience that reveals its force:

> A radical act is a confrontation with the 'root' of a problem. It is to reach for causes of an issue and not simply respond to its symptoms. It is a showdown with

limitations to embrace necessary excess and to disturb a state of affairs in pursuit of confronting those root causes... How radical performances become radical is certainly a matter of who asks and who answers the question[, Is this radical?]. (p. 18, emphasis in original)

[6] Such discourses include nuclear promotion, with policy decisions predicated on the need for meeting energy needs and mitigating climate change (Doyle, 2011), masking the inefficiencies in subsidizing nuclear (over wind, for instance) and the disasters like those in Japan (Makhijani, 2007, foreshadowing the March 2011 catastrophe).

[7] One suggestion would be to question the distinction between the source and target domain. Lakoff and Johnson (1999) explain "conflation" as how children muddy this distinction, where in adulthood, metaphors reconnect otherwise distinctive sensorimotor and subjective experiences. However, it is possible that such correlations are quite organic, and it is the material experiences of a nature–culture split that separate humans from the earth, a socialized manifestation that needlessly separate our sensorial, bodily experiences from the world we are embedded within (see Abram, 2010).

References

Abram, D. (2010). *Becoming animal: An earthly cosmology.* New York: Vintage Books.
Benford, R. D., & Hunt, S. A. (1992). Dramaturgy and social movements: The social construction and communication of power. *Sociological Inquiry, 62*(1), 36–55. doi:10.1111/j.1475-682X.1992.tb00182.x
Boston Society of Architects. (Producer). (2010, May 19). *Susannah Sayler: Canary project. Exploring design lecture series* [Podcast]. Retrieved June 22, 2012, from http://vimeo.com/12360275
Bowers, C. A. (2009). Why the George Lakoff and Mark Johnson theory of metaphor is inadequate for addressing cultural issues related to the ecological crises. *Language & Ecology, 2*(4), 1–16. Retrieved March 5, 2009, from: http://www.ecoling.net/journal.html
Charteris-Black, J. (2004). *Corpus approaches to critical metaphor analysis.* New York: Palgrave Macmillan.
Cohen, L. (2003). *A consumer's republic: The politics of mass consumption in postwar America.* New York: Knopf.
Cotter, H. (2010, May 13). Thinking green: Function over form. *New York Times,* Art Review. Retrieved June 22, 2012, from http//:www.nytimes.com.
Cozen, B. (2010). This pear is a rhetorical tool: Food imagery in energy company advertising. *Environmental Communication: A Journal of Nature and Culture, 4*(3), 355–370. doi:10.1080/17524032.2010.499212
DeLuca, K. M. (1999). *Image politics: The new rhetoric of environmental activism.* New York: Routledge.
Donoghue, J., & Fisher, A. (2008). Activism via humus: The composters decode decomponomics. *Environmental Communication: A Journal of Nature and Culture, 2*(2), 229–236. doi:10.1080/17524030802141778
Doyle, J. (2011). Acclimatizing nuclear? Climate change, nuclear power and the reframing of risk in the UK news media. *The International Communication Gazette, 73*(1–2), 107–125. doi:10.1177/1748048510386744
Elkins, J. (1998). *On pictures and the words that fail them.* Cambridge: Cambridge University Press.
Friedman, T. L. (2010). The power of green. Excerpted from T.L. Friedman, the power of green, *New York Times Magazine,* April 15, 2007. In D. Siegel & E. Morris (Eds.), *Green Patriot Posters: Images for a new activism* (pp. 12–13). New York: Metropolis Books.

Furness, Z. (2010). *One less car: Bicycling and the politics of automobility.* Philadelphia, PA: Temple University Press.

Gibbs, R. W., Jr. (2008). Metaphor and thought: The state of the art. In R. W. Gibbs, Jr. (Ed.), *The Cambridge handbook of metaphor and thought* (pp. 3–15). Cambridge: Cambridge University Press.

Green Patriot Posters. About. Retrieved October 9, 2010, from http://www.greenpatriotposters.org

Green Patriot Posters. Inspiration. Retrieved October 9, 2010, from http://www.greenpatriotposters.org

Green Patriot Posters. Posters. Retrieved October 9, 2010, from http://www.greenpatriotposters.org

Hall, D. M., Bernacchi, L. A., Milstein, T. O., & Peterson, T. R. (2009). Calling all artists: Moving climate change from my space to my place. In D. Endres, L. Sprain, & T. R. Peterson (Eds.), *Social movement to address climate change: Local steps for global action* (pp. 53–79). Amherst, NY: Cambria Press.

Hansen, A. (2011). Communication, media and environment: Towards reconnecting research on the production, content and social implications of environmental communication. *International Communication Gazette, 73*(1–2), 7–25. doi:10.1177/1748048510386739

Harold, C. (2004). Pranking rhetoric: "Culture jamming" as media activism. *Critical Studies in Media Communication, 21*(3), 189–211. doi:10.1080/0739318042000212693

Ivie, R. L. (1987). Metaphor and the rhetorical invention of cold war "idealists". *Communication Monographs, 54*(2), 165–182. doi:10.1080/03637758709390224

Jensen, A. (forthcoming). The power of urban mobility: Shaping experiences, emotions and selves on the bike. In S. Kesselring, G. Vogel, & S. Witzgall (Eds.), *Tracing the new mobilities regime.* Aldershot: Ashgate.

Kolbert, E. (2006). *Field notes from a catastrophe: Man, nature, and climate change.* New York: Bloomsbury.

Kövecses, Z. (2004). Introduction: Cultural variation in metaphor. *European Journal of English Studies, 8*(3), 263–274. doi:10.1080/1382557042000277386

Lakoff, G., & Johnson, M. (1980). *Metaphors we live by.* Chicago, IL: University of Chicago Press.

Lakoff, G., & Johnson, M. (1999). *Philosophy in the flesh: The embodied mind and its challenge to Western thought.* New York: Basic Books.

Lakoff, G., & Turner, M. (1989). *More than cool reason: A field guide to poetic metaphor.* Chicago, IL: The University of Chicago Press.

Larson, B. (2011). *Metaphors for environmental sustainability: Redefining our relationship with nature.* New Haven, CT: Yale University Press.

Lewis, J. (2010). Portraits from the edge – Kiribati – putting a face to climate change. *Visual Communication, 9*(2), 231–236. doi:10.1177/1470357209352955

Madison, D. S. (2010). *Acts of activism: Human rights as radical performance.* Cambridge: Cambridge University Press.

Makhijani, A. (2007). *Carbon-free and nuclear-free: A roadmap for U.S. energy policy.* Takoma Park, MD: IEER Press.

McGee, M. C. (2006). "Social movement:" Phenomenon or meaning? In C. E. Morris, III & S. H. Browne (Eds.), *Readings on the rhetoric of social protest* (pp. 115–26). State College, PA: Strata Pub. (Original Work Published 1980).

Metropolis. (2010, October). Patriot acts. *Metropolis: The magazine of art and design,* pp. 94–99.

Morris, E., & Sayler, S. (2008). Comment: Ruminations on the role of artists in a world of science. *Journal of Science Communication, 7*(3), 1–3. Retrieved from: http://jcom.sissa.it/archive/07/03/Jcom0703%282008%29C01/Jcom0703%282008%29C06

Nerlich, B., & Koteyko, N. (2009). Carbon reduction activism in the UK: Lexical creativity and lexical framing in the context of climate change. *Environmental Communication: A Journal of Nature and Culture, 3*(2), 206–223. doi:10.1080/17524030902928793

Ortiz, M. J. (2010). Visual rhetoric: Primary metaphors and symmetric object alignment. *Metaphor and Symbol, 25*(3), 162–180. doi:10.1080/10926488.2010.489394

Pezzullo, P. C. (2006). Articulating anti-toxic activism to "sexy" superstars: The cultural politics of *A Civil Action* and *Erin Brockovich*. In S. L. Senecah (Ed.), *Environmental communication yearbook* (Vol. 2, pp. 21–48). Mahwah, NJ: Erlbaum.

Pezzullo, P. C. (2011). Contextualizing boycotts and buycotts: The impure politics of consumer-based advocacy in an age of global ecological crises. *Communication and Critical/Cultural Studies, 8*(2), 124–145. doi:10.1080/14791420.2011.566276

Rogers, R. A. (2008). Beasts, burgers, and hummers: Meat and the crisis of masculinity in contemporary television advertisements. *Environmental Communication: A Journal of Nature and Culture, 2*(3), 281–301. doi:10.1080/17524030802390250

Shiva, V. (1990). Development as a new project of Western patriarchy. In I. Diamond & G. F. Orenstein (Eds.), *Reweaving the world: The emergence of ecofeminism* (pp. 189–220). San Francisco, CA: Sierra Club Books.

Siegel, D., & Morris, E. (2010). Destroy this book. In D. Siegel & E. Morris (Eds.), *Green Patriot Posters: Images for a new activism* (pp. 8–11). New York: Metropolis Books.

Singer, R. (2010). Neoliberal style, the American re-generation, and ecological jeremiad in Thomas Friedman's "Code Green". *Environmental Communication: A Journal of Nature and Culture, 4*(2), 135–151. doi:10.1080/17524031003775646

Sloop, J. M. (2009). People shopping. In B. A. Biesecker & J. L. Lucaites (Eds.), *Rhetoric, materiality, and politics* (pp. 67–98). New York: Peter Lang.

The Canary Project. Retrieved October 9, 2010, from http://canary-project.org/

The Canary Project. Public notice. Retrieved June 22, 2012, from http://canary-project.org/2012/04/public-notice-a-green-patriot-poster-salon/

Thill, S. (2010, November 25). *Green Patriot Posters* reinvigorate environmental message. *Wired*. Retrieved 15 June, 2012, from http://www.wired.com/underwire/2010/11/green-patriot/

Yu, N. (2008). Metaphor from body and culture. In R. W. Gibbs, Jr. (Ed.), *The Cambridge handbook of metaphor and thought* (pp. 247–261). Cambridge: Cambridge University Press.

Index

Note: Page numbers in **bold** type refer to figures
Page numbers in *italic* type refer to tables
Page numbers followed by 'n' refer to notes

Aboriginal landscape 93, 99
Acanchi 83
action: call to 34
Adorno, T. 2
advertisements 3; Chevy Silverado SUV 129; pesticides 12
advertising 69; nature 128, 129, *see also* commercials
aerial photography 29
aerial shots 9
Affliction walk-out t-shirt 126, **126**
Africa: Western cultural lens 11–12
agenda setting 36; literature 21
Agent Orange14, 41–60; bringing home 48–50; environmental policy 55; halt in application 47; images and articles on 43; liability 47, 48; natural, cultural and environmental significance 53–5; protest 50; science, uncertainty and blame 52–3; trial 50; truth and blame 52; use in Vietnam 42, 46–8; victims 48–9; victims as fathers 49
Agent Orange: Collateral Damage in Viet Nam (Griffiths) 50
Agent Orange's Bitter Harvest (Science) 53
aggression: naturalizing 128
Alberta: bituminous sands 62–3, 67; environmental reputation 67; extractive gaze 61–78; *Freedom to Create. Spirit to Achieve* brand 63; identity 65; mockery and boycott 72; prior visualizations 65; Public Affairs Bureau 66; rebranded 62–3; ties to US 71; tourist destination 63
Alberta's rebranding slideshow (*An Open Door*) 63, 66–72; images 67–71; site of audience 71–2; site of production 66–7

algal blooms 27; Chesapeake Bay 25, **26**; scientific explanation 25
Alter Ego 139
Alternative Living (poster) 157
angle: high 29; low 9, 31, 33
Anholt, S. 81
animalitarianism 131
animals: master narratives 132; objectivity 107
anthropocentricism 72–4
art: green 158
artefacts: visual 22
articulation: and metaphors 148–9; war efforts 149–51
articulation theory 147
audience: interpretation 44
Australia *see* Blue Mountains; Brand Blue Mountains

banal globalism 6
Barron, J. 49
Barthes, R. 3, 4, 5–6, 22
Baudrillard, J. 3, 4
Becker, H. 22, 24, 25
Berger, J. 3, 132
Bethlehem Steel Plant 28–9, **28**
Bhopal disaster (Union Carbide, 1984) 42, 115
bicycles 159n; images 153
biodiversity 107
Birmingham School of Cultural Studies 2
bitumen extraction, Alberta 62–3; environmental cost 63
bituminous sands: Alberta 62, 67
Blue Mountains (Australia) 14–15; photographic images 83; place branding 79–102, *see also* Brand Blue Mountains (BBM)

INDEX

Blue Mountains Tourism Limited (BMTL) 83–4
Bordo, S. 130
Bottenburg, M. van: and Heilbron, J. 133
boundaries: conservation 27
boxing 132–3
brand: definition 80
Brand Blue Mountains (BBM) 79–102; Aboriginal landscape 93, 99; Acanchi's strategy 83; applying the vision: controlling the image 97–8; background 82–3; brand graphic layout **89**, 90; brand management 97–8; brand manual 84–98; brand partners 84; color palette 87, **88**; construction 83–5; font styles 87; intelligent experiences photography 94, **95**; logo and typemark 85, **86**; macro details 96, **96**; photographic strategy 91, **92**; photographic style 90–1; standard panoramic 96–7; typography and layout 90; upness 84; vision 80; visualizing the brand: global elements 85–90; visualizing the brand: photographic strategy 90–7
branding: communicative acts 82; emotional appeals 82; experience 82; market research 81; reduction and repetition 81; uniqueness 81, *see also* place branding
Buell, L. 51
Burke, E. 91

cage fighting *see* mixed martial arts (MMA)
Canada: Fossil Fool of the Year 67, *see also* Alberta
Canary Project: Green Patriot Posters 145–62
cancer 41
capitalism 7; extractive 73, 74; green American 159n; pollution 151
caption: reading image 8
carbon: rationing 159; reduction 159
Carbon Rationing Action Groups (CRAGS) 159
Carson, R. 41, 55
causal arguments: single images 56
celebrities 7
Cerulo, K.A. 105–6
Charteris-Black, J. 147
chemical weapons 58n
Chesapeake Bay (USA) 14, 19–40; agenda setting 21; Agreement (1983) 20; algal bloom 25, **26**; geography 20; marinas and McMansions 29, **30**; natural resources 33; nature 32; Old Oak photograph **34**; role of images 22; runoff 29; sewage treatment facility 35; stakeholders 20; watershed debate 19–40
Chesapeake Bay Clean Water and Ecosystem Restoration Act (2009) 20, 21, 24, 27
Chesapeake Bay Foundation (BCF): Rapid Assessment Visual Expedition (RAVE) 19–40

Chesapeake Bay RAVE: iLCP 22–4; knowledge creation 24–7; local knowledge 26–7; outcome 35–6; scientific knowledge 24–6
Chesapeake Bay watershed construction: causes 28–31; problem 27–8; solutions 34–5; stakes 31–3
Chevy Silverado SUV advert 129
children: innocence 51
circular imagery 153–4
civilizing process 133
clean energy 156
clichés 4
climate: *Time* magazine covers 114
climate change 145–6; activism and World War II posters 147; fear-inducing imagery 14; global warming 8, 146, 155; imagery 8
coal pollution 29
color 87, **88**
commercialization: landscape 90
commercials: Nike 138–41, **138**, **140**; Tapout's *Eye of the Storm* 134–7, **135**
Commoner, B. 114, 118–19
communicative acts: branding 82
communicative context 10–11
composition 9–10; ideologies 4
compost (metaphor) 153
conjoint constitution 21
conservation: boundaries 27
Consider the Alternative (poster) 157
consumerism 62
consumption: art problematizing 151; visual 64
consumptive gaze 73
contamination 56
control is up (metaphor) 155
copyright: landscape 100
Corbett, J. 62, 64, 73, 129, 134
Corn-Oil/Water/Global Warming (poster) 154
Cosgrove, D. 90
Cottle, S. 6; and Lester, L. 7
Cousteau, J. 118
Cozen, B. 15, 145–62
crisis: masculinity 54; narrative 46
critical discourse analysis (CDA): *An Open Door* 63; visual analysis 63–4
Cronon, W. 61, 69, 73, 74; Miles, G. and Gitlin, J. 68
cultural context 11–12
cultural packages 10, 11
cultural studies: green (GCS) 126, 128
culture: notion 2–3; relationship with image 45; wild visual 125–44

Darling-Wolf, F.: and Mendelson, A. 56–7
DDT: banning 55
demand: images 33; shots 49

INDEX

development: residential 29
DiFrancesco, D.A.: and Young, N. 8, 9, 107
dioxin 42; contamination 54
dirty oil 63
discourse: nature-attuned 12; racist social-evolutionary 131
Donoghue, J.: and Fisher, A. 153
Don't be Stuck Up (poster) 152
Downey, G. 133, 134
Doyle, J. 108

East Beach (Virginia) 34, **34**
ecocentric argument 70
ecocentric perspectives 64
ecocide 41
Ecology (image) 154
economic growth 153–4; environmental art 158
ecosystems: *Time* magazine covers 115
Edgar, F.: *Affliction* walk-out t-shirt 126, **126**; snake shorts 127
Efforts Up! (poster) 156
Elevate Your Senses (BBM) 79–102
Elias, N. 133
Elkins, J. 4, 147
Emerson, R.W. 100
emotion: fact 37
emotional appeals: branding 82
emotional responses: evoking 13–14
energy: clean 156; nuclear promotion 160n; oil 62–3, 71
engagement: motivating 14
environment: visual representations 4–14
environmental affairs: defining 105
environmental art 158
environmental communication (EC) 126, 128
environmental images in mass media: literature 104
environmental issues: coding 109; mainstreaming 120; successfully constructing 36–7
environmental movements 55
environmental policy: Agent Orange 55
environmental problems: difficulties in representing 5
Environmental Protection Agency (EPA) 35
environmental spectrum 73
environmentalism: science 36
evolutionary distancing 131
extractive capitalism 73, 74
extractive gaze: Alberta 61–78

Fabian, J. 132, 141
Fable for Tomorrow (Carson) 41
fact: emotion 37
Fahmy, S. 42, 45, 50, 55

Fairey, S. 155–6
farming: pollution 29
fathers: victims as 49
Faux, W.V.: and Kim, H. 64
fear-inducing imagery 14, 34
Ferrari, M.P. 15, 46, 125–44
Field Notes from a Catastrophe (Kolbert) 146
Fisher, A.: and Donoghue, J. 153
Fitzgerald, A.: and Kalof, L. 107
Flora, C.B.: and Kroma, M.M. 12
font styles 87
football 132, 139
Foucault, M. 4
framing 44; suffering 50–1
Francis, R.D. 65
Freedom to Create. Spirit to Achieve (Alberta Government) 63
Friedman, T.L. 159, 159n
Frost, T. 24
Fryer, K. 93
future is more (metaphor) 152
future is up (metaphor) 152, 156

Gamson, W.A.: and Modigliani, A. 10, 11
garbage 29
gaze: consumptive 73; extractive 61–78; male 3; narrative 9–10; Romantic 61
Getty images 108
Gilligan, C.: and Marley, C. 44, 45
global imagery 6
global warming: imagery analysis 8; visualizing 146
Global Warming (poster) 155
globalism 62; banal 6; visualization 64
Golden Snake (image) 155
Gore, A. 119
green American capitalism 159n
green art 158
green cultural studies (GCS) 126, 128
green marketing 69
Green Patriot Posters (Canary Project) 145–62; visualized values 151–7
greenwashing 69
Griffiths, P.J. 50
Griffiths, V.J. 50
Grossfeld, S. 56
growth: economic 153–4
Guttenfelder, D. 43
Guttmann, A. 132, 139

habitus: violence 141
Hall, S. 2, 128
Hannigan, J. 36
Hansen, A. 4, 69, 128, 129, 146, 147; and Machin, D. 1–18, 42, 73, 108
Hayden, W. 130

INDEX

health 43–4
hearts and minds: winning 33
Heilbron, J.: and Van Bottenburg, M. 133
herbicides 42
heritage 64
high angle: impersonal relationships 29
historical context 12–13
History of the Future, A (Sayler) 146
Hitchens, C. 43
Hochman, J. 129
Horkheimer, M. 2
Howenstine, E. 107
human opportunity: landscapes 70
Hurricane Katrina 64, 145; Plaquemines Parish 146

iconic imagery 9
identity 64, 65, 100
ideologies: composition 4
imagery: circular 153–4; fear-inducing 14, 34; global 6; iconic 9; nature 127, 129; protest 50; shocking 28, 37; symbolic 126; toxicity and health understanding 43–4
images 22; Agent Orange 43; Alberta's slideshow (*An Open Door*) 67–71; bicycles 153; Blue Mountains (Australia) 83; demand 33; *Ecology* 154; environmental 104; Getty 108; *Lumberjack* 154; offer 33; *Paint it White* 157; polysemic 44; single 56; social effects 64; speed in processing 44, *see also* photographs; toxic images
imperialism 55
impersonal relationships 29
innocence: children 51
International League of Conservation Photographers (iLCP) 14, 19–40; mission 23
interpretation: audience 44; ready-made 45
Ivie, R.L. 158

Jantzen, G. 141–2
Jeffords, F. 50
Jenks, C. 3
Joffe, H.: and Smith, N.W. 8, 107
Johnson, M.: and Lakoff, G. 157, 160n
Join the Revolution (poster) 154

Kalof, L.: and Fitzgerald, A. 107
Keep Buying Shit (poster) 152
Kim, H.: and Faux, W.V. 64
Kiribati 145, 158
knowing 3, 5
knowledge: creation 24–7; local 26–7
Kolbert, E. 146
Kosovo photographs 45
Koteyko, N.: and Nerlich, B. 158, 159
Kozol, W. 45

Kress, G.: and van Leeuwen, T. 4, 6, 33
Kroma, M.M.: and Flora, C.B. 12
Kurasawa, F. 131, 139

Lakoff, G.: and Johnson, M. 157, 160n
landscape: commercialization 90; copyright 100; human opportunity 70; visual consumption 64
Lange, D. 51
legacies: Vietnam War 42
Lester, L.: and Cottle, S. 7
Linden, E. 104
linear growth 153
livestock 25
local knowledge 26–7
logos 85–7
Loi, R. 120
Love Canal (NY): waste dump 42, 57n
low angles 9; disempowerment 31; identification 33
Lumberjack (image) 154

McGeachy, L. 120
Machin, D.: and Hansen, A. 1–18, 42, 73, 108
McQuail, D. 104
made to mean visuals (Hansen) 146, 147
Madison, D.S. 159–60n
magazine covers: studies 105, *see also Time* magazine covers
male animal ideology 128, 130, 139
male gaze 3
manipulation software: Photoshop 22
Manzo, K. 108
marinas and McMansions (Chesapeake Bay) 29, **30**
marketing: green 69; tourism 80
Marley, C.: and Gilligan, C. 44, 45
Marx, K. 142, 159n
masculine primitive 128, 130–2
masculinity crisis 54
master narratives 132; Western 141
Mauss, M. 134
meaning 64; fixing 45
meaning-making 44; sites 13–14
media 104; products 3
Meisner, M.S. 118; and Takahashi, B. 15, 103–24
Mellor, F. 8
Mendelson, A.: and Darling-Wolf, F. 56–7
metaphors 152, 153, 155, 156, 157; and articulation 148–9; more is better 151, 153, 155–7; more is up 151–3; orientational 147
metaphysical nature 141–2
minority victims 64
Mirzoeff, N. 64
mitigation techniques 34, 35

INDEX

mixed martial arts (MMA) 15; de-sportizing 133; nature imagery 127; spectacle 133–4; t-shirt 136–7; wild visual culture 125–44
Modigliani, A.: and Gamson, W.A. 10, 11
Monsanto Chemical Company 47
more is better (metaphor) 151, 153, 157; up over down 155–7
more is up (metaphor) 151–3
Morris, E. 146
Morton, T. 61, 63
mothers 51
movement: social 154
Mugs are Great (poster) 152, 154
Mulvey, L. 3
myths 4

Nachtwey, J. 43, 44, 51, 52
Nader, R. 119
Nansemond River 31, **32**
narrative: crisis 46; gaze 9–10; master 132, 141; visual 44–5, 46, 56
National Geographic: studies 107
natural: deconstructing 128–9
natural resources 33
naturalizing aggression 128
nature: advertising 128; as backdrop 141; Chesapeake Bay 32; commodifying 141–2; cultural production 128–30; defining 105; iconographies 126; imagery 127, 129; master narratives 132; metaphysical 141–2; primitivity 125–44; selling 61–78; separated from people 69, 73; serving people 70; sporting 134–41; subjective construction 74; symbolic domestication 118; as threat 140; use in advertising 128, 129
nature-attuned discourse 12
neoliberalism 72
Nerlich, B.: and Koteyko, N. 158, 159
New York City: future water-line cutting 146
Nicholson-Cole, S.A. 107
Nike commercials: *Alter Ego* 138–9, **138**; *For Warriors Only* 139–41, **140**
Noble, A. 53, 54
nostalgia 12, 32
nuclear promotion 160n

Oates, J.C. 132, 133
Obama: Hope (posters - Fairey) 156
Obama, B. 71
objective reporting 45
Occupy Movement 73
offer images 33
oil 62–3; consumption (USA) 71; dirty 63
On Boxing (Oates) 132, 133
Open Door, An (Alberta slideshow) 63, 66–72; critical discourse analysis (CDA) 63

Operation Hades 46
Operation Ranch Hand 46
opportunity: human 70
orientational metaphors: up-down 147
orientations 147
Ortiz, M.J. 151

Paint it White (image) 157
Pauwels, L. 2
Peeples, J. 14, 41–60
people: disconnected with environment 70; separated from nature 69, 73
perspective 9–10
pesticides: advertisements 12
Peterson, A. 138
Pew Research Center: Project for Excellence in Journalism 104
Pezzullo, P.C. 157
photographers: iLCP 14, 19–40
photographs: Kosovo 45; social movements 22
photography: aerial 29; Brand Blue Mountains (BBM) 90–1, **92**; power 3; society functions 22; trust 44
photojournalism 45, 53
Photoshop software 22
Pictures of the Year International (POYI) 43
place branding 14, 64, 80–2; Blue Mountains 79–102; strategy 79; tourism 69
Plaquemines Parish: Hurricane Katrina 146
point of view 9–10
pollution: capitalism 151; Chesapeake Bay 27, 28, 29; coal 29; farming 29; *Time* magazine covers 114
polysemic images 44
Porter, N. 14–15, 79–102
posters 151–7; Fairey 155–6; Green Patriot Posters (Canary Project) 145–62
power 55, 74; photography 3; solutions 37
Power Up Windmill (Fairey) 155–6
Power Up Windmill poster 155–6
primitive myth 130–1
primitivism: socio-psychological 131
primitivity: nature 125–44
Pro Combat 138
Promised Land 65
protest: Agent Orange 50; imagery 50

QDA Miner software 109

racism 55
racist social-evolutionary discourses 131
radical act 159–60n
reading image: caption 8
Reading Images (Kress and van Leeuwen) 4, 6, 33
real: represented 3

INDEX

Recycle poster 154
relationships: impersonal 29
Remillard, C. 107
reporting: objective 45
representations: environment 4–14
represented: real 3
research: visual environmental communication 1–18
residential development 29
resource-based economy: selling nature 61–78
risk: toxic 45–6, 50–2; visual narratives 46
Robinson, W. 91, 93, 99
Romantic gaze 61
Rose, G. 13, 66, 73–4
rugby 132
runoff 29; Chesapeake Bay 29; mitigation 34

Saussure, F. de 8
Sayler, S. 146
Schecter, A. 53
Schoenfeld, A.C., Meier, R.F. and Griffin, R.J. 5
Schwarz, E. 14, 19–40
science: discourse 12; environmentalism 36
scientific knowledge: Chesapeake Bay RAVE 24–6
sea levels: rising 27
Seager, J. 43
seeing 3, 5; new way 22
selling nature: resource-based economy 61–78
semiology 8–9
Seppänen, J.: and Väliverronen 107
sewage treatment facility: Chesapeake Bay 35
shocking imagery 28, 37
Shuck, P. 47
sign systems: interacting 8–9
Silent Spring (Carson) 55
single images: causal arguments 56
Smith, J. 87
Smith, N.W.: and Joffe, H. 8, 107
social effects: images 64
social history 132
social modality 66
social movements 37, 154; photographs 22
societal processes 72–4
socio-psychological primitivism 131
software 22, 109
solutions 34; power 37
Sontag, S. 3
Soper, K. 128
S.O.S. (poster) 152–3
sporting nature 134–41
sportization 133
sports studies: recourse to primitive and natural 132–4
Stone, D.A. 21, 22, 36
suffering 56, 115; framing 50–1

Sumner, D.E. 105
supremacy: white 136
Sustain (poster) 153
sustainability 12; *Time* magazine covers 119; visualizing 146
symbolic imagery 126
Szerszynski, B. 6, 7

Tagg, J. 65
Takach, G. 14, 61–78
Takahashi, B.: and Meisner, M.S. 15, 103–24
talking heads: wall-paper shots 8
Tapout: *Eye of the Storm* advert 134–7, **135**
testimony: expressing suffering 56
think differently 158
Time magazine 15; circulation 104
Time magazine covers: action 119; animals 114; aspects of nature represented 116; categories of representations of nature 116, *118*; climate 114; Commoner 114, 118–19; Cousteau 118; covering environments 105–8; as cultural artefact 105; defining environmental affairs 105; ecosystems 115; energy efficiency 119; environmental affairs 103–24; fragmentation of issues 114; *Global Warming* 115; Gore 119; green design 119; *How to Save the Earth* 104; *how to win the war on global warming* 119; human suffering 115; issues representation 114–16; issues represented 110, *111–13*; list of issues 124; nature representation 116, 118; numbers with environmental issues 110, **110**; *Our Filthy Seas* 115; people 118–19; *Planet of the Year* 104; pollution 114; studies on 105–7; study data analysis 109–10; study data collection 108–9; study questions 108; sustainability 119; text 114; types of aspects of nature 116, *117*; unintended human consequences 114; *Vanishing Act* 116
Todd, A.M. 69, 107
Torgovnick, M. 131, 136
tourism: Alberta as destination 63; marketing 80; place branding 69
toxic images 43–4, 55; cultural and natural contexts 55; lessening uncertainty 57; obfuscating complexity 56–7; victims 48–9; victims as fathers 49; visual narrative 56
toxic risk: images 45–6; Vietnamization 50–2
toxicity and health 43–4
toxins 14; imaging 41–60; invisibility 45
Tracking Agent Orange (Watriss) 49
trust: photography 44

Ultimate Fighting Championship (UFC) 126–7, 133–4, 136, 142
Union Carbide: Bhopal disaster (1984) 42, 115

INDEX

United States of America (USA): New York City 146; oil consumption 71, *see also* Chesapeake Bay
up is good (metaphor) 152
up over down: more is better (metaphor) 155–7
up-down: orientational metaphors 147
up-down binary: alternative conceptions 153–5
Upside Down (poster) 151

Väliverronen: and Seppänen, J. 107
values: visualized and Green Patriot Posters 151–7
Vanity Fair 43
Venum brand 127
victims 48–9; as fathers 49; minority 64; representation 50
Vietnam Inc (Griffiths) 50
Vietnam War *see* Agent Orange
Vietnamization: toxic risk 50–2
violence: habitus 141; natural notion 126; romantically naturalizing masculine 142
visual analysis: contexts and sites 10; critical discourse analysis (CDA) 63–4

visual communication: how it works 4; researching 2–4

wall-paper shots: talking heads 8
war efforts: articulating 149–51
Water's Rising (poster) 152–3
Watriss, W. 49
Waugh, C. 41, 43, 55
Ways of Seeing (Berger) 3
Western cultural lens: Africa 11–12
Western master narratives 141
White, D. 142
white supremacy 136
Williams, R. 128
Winds of Change: Climate, Weather and the Destruction of Civilizations, The (Linden) 104
winning hearts and minds 33
witness 44
words: processing time 44
World War II posters: climate change activism 147

Yannacone, V.J. 50
Young, N.: and DiFrancesco, D.A. 8, 9, 107

For Product Safety Concerns and Information please contact our EU representative GPSR@taylorandfrancis.com Taylor & Francis Verlag GmbH, Kaufingerstraße 24, 80331 München, Germany

Printed and bound by CPI Group (UK) Ltd, Croydon, CR0 4YY
08/06/2025
01896999-0017